AQA Mathematics

for GCSE

Exclusively endorsed and approved by AQA

Series Editor
Paul Metcalf

Series Advisor
David Hodgson

Lead Author
Steven Lomax

June Haighton
Anne Haworth
Janice Johns
Andrew Manning
Kathryn Scott
Chris Sherrington
Margaret Thornton
Mark Willis

HIGHER
Linear 2

Nelson Thornes
a Wolters Kluwer business

Published in 2006 by:
Nelson Thornes Ltd
Delta Place
27 Bath Road
CHELTENHAM
GL53 7TH
United Kingdom

06 07 08 09 10 / 10 9 8 7 6 5 4 3 2 1

A catalogue record for this book is available from the British Library.

ISBN 0 7487 9778 5

Cover photograph: Seals by Stephen Frink/Digital Vision LU (NT)
Page make-up by MCS Publishing Services Ltd, Salisbury, Wiltshire

Printed and bound in Spain by GraphyCems

Acknowledgements

The authors and publishers wish to thank the following for their contribution:
David Bowles for providing the Assess questions
David Hodgson for reviewing draft manuscripts

Thank you to the following schools:
Little Heath School, Reading
The Kingswinford School, Dudley
Thorne Grammar School, Doncaster

The publishers have made every effort to contact copyright holders but apologise if any have been overlooked.

Contents

Introduction

This book contains homework that allows you to practise what you have just learned. Each chapter is divided into sections that correspond to the numbered Learn topics for the matching chapter in the Students' Book.

 Means that these questions should be attempted with a calculator.

Means that these questions are practice for the non-calculator paper in the exam and should be attempted without a calculator.

1 ← Underlined questions are harder questions.

Coursework

This section explains the coursework mark scheme and features three Using and applying mathematics mini-coursework tasks.

1 Properties of circles

Homework 1

1 Calculate the marked angles in these diagrams, giving reasons for your answers. The centre of the circle is marked O.

a

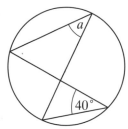

40°, a

d

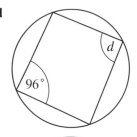

96°, d

f

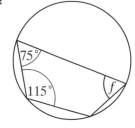

75°, 115°, f

b

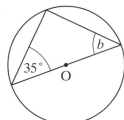

35°, O, b

e

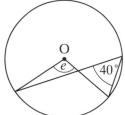

O, e, 40°

g

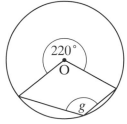

220°, O, g

c

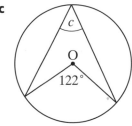

c, O, 122°

2 The chord AB subtends an angle of 104° at the centre of the circle.

Calculate the angles ADB and ACB, giving reasons for your answers.

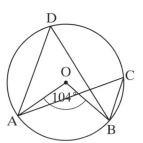

D, O, C, 104°, A, B

3 AD is a diameter of the circle, centre O.

Angle ADC = 70° and angle ACB = 25°.

Calculate:

a angle DCA

b angle DAC

c angle ABC

d angle DAB.

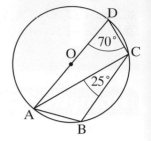

4 Sarah says that a rhombus cannot be a cyclic quadrilateral.

Is she correct?

Give a reason for your answer.

5 The lines PR and QS pass through the centre of the circle at O.

Angle QRP = 32°.

Calculate:

a angle QSP

b angle RQS

c angle RPS

d angle POS.

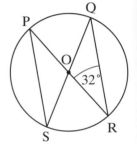

6 AC is a diameter of the circle, centre O.

Angle BOC = 106° and angle ACD = 35°.

Calculate:

a angle CAB

b angle ADC

c angle DAC

d angle DAB.

Give reasons for your answers.

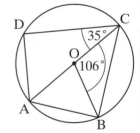

7 A quadrilateral is drawn inside a circle, centre O.

Angle OPQ = 50° and angle POR = 165°.

Calculate the remaining angles of the quadrilateral.

Remember to show all your working.

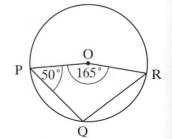

Homework 2

1 Calculate the marked angles in these diagrams, giving reasons for your answers.

The centre of the circle is marked O.

a

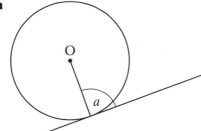

e

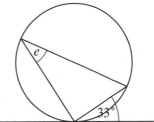

b

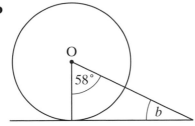

f

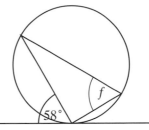

c

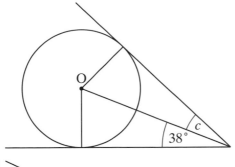

g

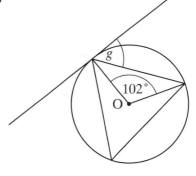

d

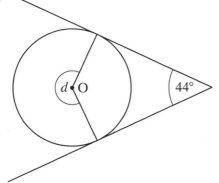

2 PR is a tangent to the circle, centre O.

The tangent touches the circle at Q.

Angle QOS = 142°.

Calculate the angle QRO.

Remember to show all your working.

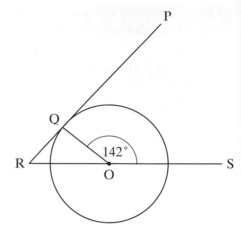

3 PQ is a tangent to the circle.

Angle PAC = 34°.

Calculate the angle CBA.

Give a reason for your answer.

4 PA is a tangent to the circle, centre O.

PB is a straight line passing through O.

Angle PAC = 32°.

Calculate the angles CBA and CPA.

Remember to show all your working.

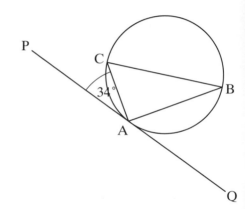

5 PQ is a tangent to the circle, centre O.

Angle COA = 88°.

Calculate the angle CBA and the angle CAP.

Remember to show all your working.

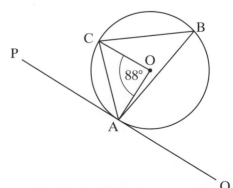

6 Tangents SP and SR meet the circle at P and R respectively.

SOQ is a straight line passing through the centre of the circle, O.

Angle PSR = 50°.

Calculate:

a angle PSO

b angle SPO

c angle POS

d angle POR

e angle PQR.

Give reasons for your answers.

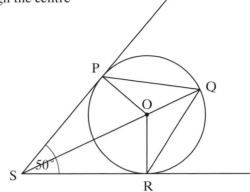

7 XY is a tangent to the circle touching at the point A.

AB = BC and CD = DA.

Angle XAB = x.

Calculate, leaving your answers in terms of x:

a angle BCA

b angle ABC.

Angle ADC = 96°.

c Find the value of x.

Give a reason for your answer.

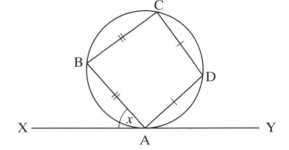

8 CD is the tangent to the circle at C.

Calculate the value of b.

Give reasons for your answer.

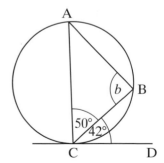

9 Prove that the angle CAP is the same as angle ABC.

Remember to show all your working.

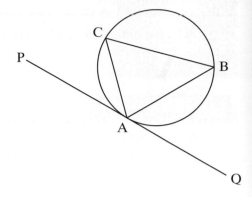

2 Trial and improvement

Homework 1

1 Use the trial and improvement method to solve these equations.

Give your answers to one decimal place.

Remember to show all your working.

   a $x^2 = 72$ **b** $x^2 = 44$ **c** $x^3 = 105$ **d** $x^3 = 54$

2 The area of this square is 110 cm^2.

Area
110 cm^2

Use the trial and improvement method to find the length of the square.

Give your answer to two decimal places.

3 The volume of a dice is 4 cm^3.

Use the trial and improvement method to find the length of a side of the dice.

Give your answer to two decimal places.

4 Use the trial and improvement method to find the length of this rectangle to two decimal places.

Remember to show all your working.

$(x+4)$

x | Area $= 90 \text{ cm}^2$

5 Use the trial and improvement method to find the length of this rectangle to two decimal places.

Remember to show all your working.

$2x$

$(x-4)$ | Area $= 55 \text{m}^2$

6 The area of Amir's garden is 100 square metres.

The length of the garden is two metres longer than the width.

Use the trial and improvement method to work out the length and width of the garden.

Give your answers to one decimal place.

7 A solution of the equation $x^3 - 8x = 110$ lies between $x = 5$ and $x = 6$.

Use the trial and improvement method to find this solution.

Give your answer to one decimal place.

8 Use the trial and improvement method to find solutions to these equations.

Give your answers to two decimal places.

a $x^2 + 3x = 46$ if the solution lies between 5 and 6

b $x^3 + 2x = 38$ if the solution lies between 3 and 4

c $x^3 + 21 = 0$ if the solution lies between -2 and -3

d $x^3 + 2x = -78$ if the solution lies between -4 and -5

1 Write the coordinates of the images of these points after the translations described.

    a (2, 4) translation 3 units to the right followed by 1 unit up

    b (3, 6) translation 4 units to the right followed by 6 units down

    c (−1, 0) translation 3 units to the right followed by 2 units up

    d (−1, 3) translation 3 units to the left followed by 4 units down

    e (−2, −9) translation 3 units to the left followed by 4 units up

2 Find the coordinates of the image of each of these points after a translation by the given vector.

    a (3, 4) by $\begin{pmatrix} 11 \\ 7 \end{pmatrix}$ **c** (0, 4) by $\begin{pmatrix} 7 \\ -8 \end{pmatrix}$

    b (2, 0) by $\begin{pmatrix} 3 \\ 4 \end{pmatrix}$ **d** (−2, −3) by $\begin{pmatrix} -5 \\ -4 \end{pmatrix}$

3 Draw axes for both x and y between 0 and 10.

    Draw triangle A, with vertices at (1, 3), (1, 7) and (3, 3).

    a Draw the image of triangle A after the translation $\begin{pmatrix} 2 \\ 3 \end{pmatrix}$. Label the image B.

    b Draw the image of triangle A after the translation $\begin{pmatrix} 3 \\ 2 \end{pmatrix}$. Label the image C.

    c Describe the translation that would move triangle B onto triangle C.

4 Draw axes for both x and y between 0 and 6.

    Draw triangle X, with vertices at (2, 3), (4, 5) and (5, 1).

    a Draw the image of triangle X after the translation $\begin{pmatrix} -2 \\ -1 \end{pmatrix}$. Label the image Y.

    b Draw the image of triangle Y after the translation $\begin{pmatrix} 0 \\ 2 \end{pmatrix}$. Label the image Z.

    c What are the coordinates of the vertices of triangle Z?

5 Triangle ABC is translated onto triangle A′B′C′. A and B are the points (2, 2) and (7, 2) respectively. A′ and C′ are the points (−3, −5) and (−3, −1) respectively.

Find the coordinates of C and B′.

6 Draw axes from −6 to 6.

Draw the quadrilateral A, with coordinates (−1, 3), (1, 3), (1, 6) and (−1, 6).

a Draw the image of this quadrilateral after the translation $\begin{pmatrix} -4 \\ -2 \end{pmatrix}$.

Label the image B.

b Draw the image of B after the translation $\begin{pmatrix} 2 \\ -6 \end{pmatrix}$. Label the image C.

c What vector would translate C back to A?

Homework 2

1 Draw axes for x between −4 and 8 and y between −2 and 5.

Plot the quadrilateral W using the coordinates (3, 2), (4, 2), (4, 4), and (3, 4).

a Translate the quadrilateral W by the vector $\begin{pmatrix} 4 \\ 0 \end{pmatrix}$. Label the image X.

b Reflect X in the line $x = 2$. Label this image Y.

c Describe the single transformation that maps W directly onto Y.

2 Draw the axes for x between −4 and 8 and y between −2 and 5.

Plot the quadrilateral Z, with coordinates (3, 2), (4, 2), (4, 4) and (3, 4).

a Reflect Z in the line $x = 2$. Label the image A.

b Translate the quadrilateral A using the vector $\begin{pmatrix} 4 \\ 0 \end{pmatrix}$. Label the image B.

c Describe the single transformation that maps Z directly onto B.

3 Draw axes from −6 to 6.

Draw triangle T with vertices at (1, 1), (4, 1) and (4, 5).

a Reflect T in the *x*-axis and label the image U.

b Translate U using the vector $\begin{pmatrix} -5 \\ -1 \end{pmatrix}$ and label the image V.

c Reflect V in the *x*-axis and label the image W.

d Describe the single transformation that maps T directly onto W.

4 Draw the *x*-axis from −9 to 9 and the *y*-axis from −5 to 5.

Plot triangle L whose vertices are the points (2, 1), (5, 1) and (5, 4).

a Reflect L in the *y*-axis and label the image M.

b Reflect M in the *x*-axis and label the image N.

c Rotate L through 180°, with centre (5, 1), and label the image P.

d What single transformation maps L onto N?

e What single transformation maps N onto P?

Homework 3

1 **a** Copy and enlarge these shapes, making every line twice as long.

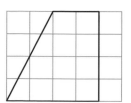

b Work out the areas of the original and the enlarged shapes.

c What do you notice?

2 Copy and enlarge this shape by scale factor:

a 3 **b** $\frac{1}{2}$ **c** 1.5

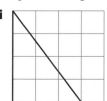

3 This rectangle is enlarged with a scale factor 2.5

What are the measurements of the enlarged rectangle?

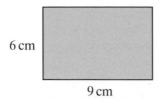

6 cm

9 cm

4 A rectangle has been enlarged. The original rectangle has length 20 cm. The enlarged rectangle has length 15 cm.

   a What is the scale factor of enlargement?

   b If the width of the enlarged rectangle is 12 cm, what is the width of the original rectangle?

   c Find the areas of the two rectangles.

5 Rectangle L maps onto rectangle M under an enlargement, scale factor 2. The width of L is 2 cm and the length is 3 cm.

What is the area of rectangle M?

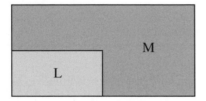

M

L

Homework 4

1 For each part, copy the diagram onto squared paper and draw an enlargement with scale factor 2.

Use C as the centre of enlargement.

a

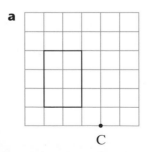

C

b

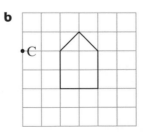

•C

2 For each part, copy the diagram onto squared paper and draw an enlargement with scale factor $\frac{1}{2}$

Use C as the centre of enlargement.

a

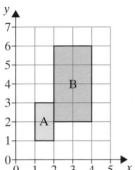

b

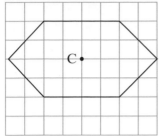

3 In each diagram below, shape A has been enlarged to make shape B.

In each case find the scale factor and the centre of the enlargement.

a

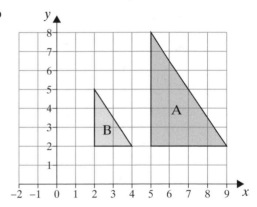

b

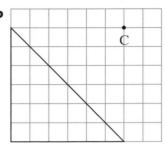

4 Draw axes from −8 to 8.

Draw triangle D using the coordinates (1, −2), (7, −2) and (−3, 8).

a Draw the image of D after an enlargement with scale factor $\frac{1}{2}$, centre (−5, −8). Label the image E.

b What are the coordinates of the vertices of E?

5 Draw axes from −6 to 6.

Draw triangle G using the coordinates (1, 1), (3, 3) and (3, 2).

a Draw the image of G after an enlargement with scale factor −2, centre (1, 0). Label the image H.

b What are the coordinates of the vertices of H?

Homework 5 ▦

1 Triangle B is an enlargement of triangle A.

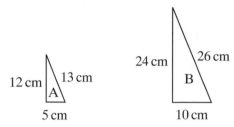

a What is the scale factor of the enlargement?

b What is the ratio of the lengths of their bases?

c What is the ratio of their longest sides?

d What is the ratio of the perimeters?

e What is the ratio of the areas of the triangles?

2 Triangle CDE is an enlargement of triangle FGH.

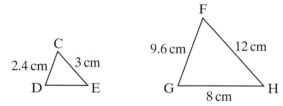

a What is the scale factor of the enlargement?

b What is the length of DE?

c What is the ratio of the perimeters of these triangles (in its simplest form)?

d What is the ratio of the areas of the triangles?

3 Rebecca says that triangle K is an enlargement of triangle J.
Is she correct? Give a reason for your answer.

4 Triangle Q is an enlargement of triangle P.

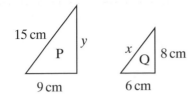

15 cm   *y*   *x*   8 cm
P   Q
9 cm   6 cm

    a What is the ratio of the corresponding sides?

    b What are the lengths of the sides x and y?

    c What is the ratio of the perimeters of these triangles?

    d What is the ratio of the areas of these triangles?

5 Two similar cylindrical cans hold 0.5 litre and 4 litres of oil. The smaller can has a radius of 6 cm.

    a What is the ratio of the volumes?

    b What is the radius of the larger can?

6 A rectangular baking tin has a base area of 60 cm^2 and a volume of 300 cm^3. A similar tin has a base area of 240 cm^2.

    What is the volume of the larger tin?

Homework 6

1 State whether each of these quantities is a length, an area or a volume.

    a Diameter of a circle   **d** 63 m^2

    b Amount of drink in a can   **e** 4 cm

    c Space covered by a square   **f** 33 mm^3

2 The letters a, b, c, x, y, z each represent a length. State for each of these expressions whether they could represent length, area or volume.

    a a^2 **c** cx **e** πab **g** $4a^2z$

    b $2b$ **d** πyz **f** $\dfrac{xy}{2}$ **h** $(b+c)^2$

3 James wanted to work out the volume of a certain solid. He looked up in his notes to find the formula. He found the formula but the ink was smudged so he could not read the index number. The formula read $V = \frac{4}{3}\pi r^{\blacklozenge}$. What was the index number?

4   For these formulae, decide whether Y could represent a length, an area, a volume or makes no sense.

a, b, c, x, z each represent a length.

a $Y = 2ab + z^2$

c $Y = 3\pi a^3$

b $Y = \dfrac{abc}{xz}$

d $Y = (b + c)^2 + z^3$

5   In these formulae A represents area, V represents volume and x, y, z are all lengths. State whether these formulae could represent length, area, volume or makes no sense.

a $\dfrac{Ax}{V}$

b $\pi x + \pi y^2$

c $x(A - y^2)$

d $A(x + y)$

6   From this list of expressions write those that could represent:

a length   **b** area   **c** volume   **d** none of these.

a, b, c and d all represent length.

i $2\pi a^2 + 3\pi cd$

iii $\dfrac{(a + b)^2}{d}$

v $a(a + b)(c + d)$

ii $ab(c + d)$

iv $\pi^2(c + d)$

vi $\pi abc + d$

4 Measures

1   **a** A swimmer swam 150 m in 5 minutes.
      What was her average speed in metres per second (m/s)?

   b A spider runs at a steady speed of 4 cm/s for 7.5 seconds.
      How far does it travel?

   c How long does it take a motorcyclist to travel 80 miles at an average speed of
      64 mph? (Give your answer in hours and minutes.)

   d Viv skis for 15 minutes at a steady speed of 12 mph.
      How far does she travel?

   e Concorde travelled 5600 kilometres from London to New York in 3 hours
      and 20 minutes.

      Calculate its average speed:

      i in km/h   **ii** in mph

   f How long does it take biscuits to move along a 10 m section of a production
      line at a steady speed of 2 cm/s?

2 **Get Real!**

   The diagram gives the times that a train stops at stations on a journey
   from Cardiff to Derby and the distances between these stations.

45 miles	43 miles	50 miles	41 miles

Cardiff	Bristol	Cheltenham	Birmingham	Derby
09:05	09:58	10:42	11:26	12:11

   Find the average speed:

   a between each station and the next

   b for the whole journey.

3   An ambulance sets off at 10:35 a.m. to pick up a patient who lives
      12 miles from the hospital. Its average speed on the journey is 40 mph.

   a At what time does the ambulance pick up the patient?

   The ambulance immediately returns to the hospital, arriving there
   at 11:17 a.m.

   b What is its average speed:

      i on the return journey   **ii** for the whole journey?

4 Get Real!

The table gives some of the results from women's athletics in the 2004 Olympic Games.

Winner	Distance	Time
Yuliya Nesterenko	100 m	10.93 s
Veronica Campbell	200 m	22.05 s
Tonique Williams-Darling	400 m	49.41 s
Kelly Holmes	800 m	1 min 56.38 s
Kelly Holmes	1500 m	3 min 57.90 s
Meseret Defar	5000 m	14 min 45.65 s
Xing Huina	10000 m	30 min 24.36 s

Find the average speed of each runner in:

a metres per second **b** kilometres per hour.

5 Five students are told that an insect flies 1.5 kilometres in half an hour.

The answers they give for the average speed are:

Kelly 45 km/h   Lee 1.2 m/s   Steve 50 m/s

Tom 20 cm/s   Meena 83 cm/s

a Who gave the correct answer?

b Explain what you think each of the other students did wrong.

6 The average speed for a journey was 5 metres per second.

a Give a possible distance in metres and time in seconds.

b Give a possible distance in kilometres and time in minutes.

7 A motorcyclist starts a journey of 225 miles at 1 p.m.

a He expects to travel at an average speed of 50 mph.
What time does he expect to arrive?

b If his motorcycle's rate of petrol consumption is 15 miles per litre, how much petrol will he use:

   i   in litres   **ii** in gallons (to the nearest gallon).

8 Get Real!

   a i   Find the distance you travel in a journey that you do regularly.

     ii Time how long this journey takes on a number of occasions.

b Work out your average speed on each journey and your overall average speed.

9 Get Real!

The table gives the radius of the equator of each planet in the solar system and the time it takes to rotate once on its axis.

Planet	Mean radius (km)	Time
Mercury	2439	58.65 days
Venus	6051	243.0 days
Earth	6378	23 hours 56 minutes
Mars	3397	24 hours 37 minutes
Jupiter	71 492	9 hours 55 minutes
Saturn	60 268	10 hours 39 minutes
Uranus	25 559	17 hours 14 minutes
Neptune	24 760	15 hours
Pluto	1123	6.39 days

Find the speed of a point on the equator of each planet in metres per second to 3 significant figures.

 10 Copy and complete the table.

Mass	Volume	Density
745 g	50 cm³	
6.4 kg	0.8 m³	
	24 cm³	3.5 g/cm³
	1.3 m³	560 kg/m³
342 g		0.9 g/cm³
3.6 kg		450 kg/m³

 11 The density of hydrochloric acid is 1200 kg/m³.
What mass of acid would fill a 20 litre tank?

Give your answer:

a in kilograms   **b** in pounds.

 12 A cylindrical concrete post weighs 3.6 tonnes.
It is 3 m long and has a diameter of 80 cm.

Find its density in g/cm³.

13 The density of copper is 8.9 g/cm³ and the density of zinc is 7.1 g/cm³.

An alloy is made by melting and mixing 800 g of copper and 200 g of zinc.

Find the density of this alloy.

14 Find the value of the quantity named in each part.

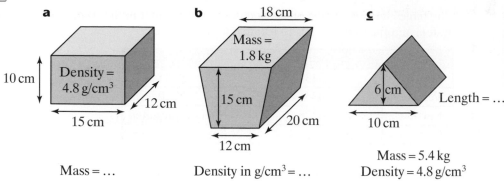

a

10 cm

Density = 4.8 g/cm³

15 cm

12 cm

Mass = …

b

18 cm

Mass = 1.8 kg

15 cm

12 cm

20 cm

Density in g/cm³ = …

c

6 cm

10 cm

Length = …

Mass = 5.4 kg
Density = 4.8 g/cm³

15 Venus has a mass 4.87×10^{21} tonnes and a volume of 9.28×10^{11} km³.

Find its average density in kg/m³.

16 A sports trophy consists of a triangular glass prism mounted on a wooden block.

The dimensions are shown in the diagram.

The density of the glass is 2.6 g/cm³ and the density of the wood is 0.58 g/cm³.

How much does the trophy weigh?

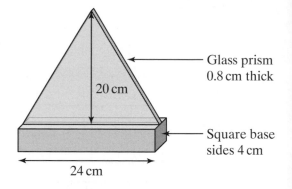

Glass prism 0.8 cm thick

20 cm

Square base sides 4 cm

24 cm

17 Population density is the number of people per square kilometre who live in an area.

a An inner city area covers an area of 12 square kilometres and has a population density of 3545 people per square kilometre. How many people live in the inner city area?

b The outer areas of the city cover 36 square kilometres and have a population of 32 472. What is the population density in this part of the city?

c Calculate the population density of the whole city.

Homework 2

 1 Calculate the maximum possible average speed for a journey of 53 miles in 1.5 hours, where both measurements are given correct to two significant figures.

 2 Allan drives 27.8 miles in 38.5 minutes. Calculate Allan's maximum possible average speed if both values are correct to 3 significant figures.

 3 Tasmin cycled 18.6 kilometres to her friend's house. She travelled a different route on her way home and found the distance was 17.9 kilometres.

What is the minimum difference in kilometres between the two routes?

 4 Get Real!

Mount Everest is 8850 m high and Mount Kangchenjunga is 8585 m high. If both heights are given to the nearest 5 metres, what is the greatest possible difference between them?

 5 A bag of sugar contains 1 kg to the nearest 5 g. If it is divided equally between five identical bowls, write the amount in each bowl in the form $a \leqslant$ amount in each bowl $< b$.

 6 A caterer estimates that a serving of soup is 200 mℓ to the nearest 50 mℓ.

Find upper and lower bounds for the total amount needed for 80 servings.

 7 A teacher gives her students this question:

'A roll of ribbon is 4 metres long to the nearest 10 cm. It is cut into pieces that are each 15 cm long to the nearest centimetre. Find the maximum possible number of pieces.'

The working of three students is given below.

Seeta's working	Kate's working	Greg's working
$405 \div 15.5 = 26.12...$	$405 \div 14.5 = 27.93...$	$405 \div 14.5 = 27.93...$
Answer: 26 pieces	Answer: 27 pieces	Answer: 28 pieces

  a Who is correct?

  b Explain the mistakes each of the others has made.

8 The radius of a circle is 3.4 cm correct to two significant figures.

    a Find the upper and lower bounds of the circumference of the circle.

    b What difference does it make if you use 3.14 instead of the π key?

 9 x, y and z are three measurements. In each case, say whether you will need to use the maximum possible or the minimum possible value of each measurement in order to find:

    a maximum value of $x + y - z$

    b minimum value of $x + y - z$

    c maximum value of $\dfrac{x}{y - z}$

 10 A metal bar has a volume of 125 cm³, correct to the nearest cm³, and weighs 2405 g correct to the nearest gram. Pure gold has a density of 19.4 g/cm³.

    Calculate the upper and lower bounds of the density of the metal bar and consider whether it could be pure gold.

 11 A kite has diagonals of length 48 cm and 36 cm, correct to the nearest centimetre.

    What is the minimum possible area of the kite?

Homework 1

1 Mo the courier delivers meals-on-wheels from her depot in Yumville.
She leaves the depot at 10:45 a.m.

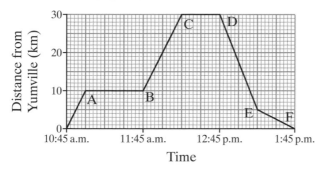

a How many times does Mo stop to deliver meals?

b Calculate Mo's speed during section OA in kilometres per hour (km/h).

c Calculate Mo's average speed over the first $1\frac{1}{2}$ hours of the journey.

d If Mo exceeds 25 mph she spills the soup. Does Mo spill the soup?

e Calculate Mo's average speed for the whole journey.

2

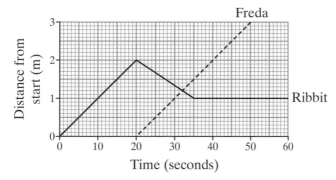

Write a commentary for the Three-metre Frog Race between Freda
and Ribbit.

You should include key distances, times and speeds in your commentary.

3 Luigi leaves his house at 9:30 a.m. and cycles to the post office, which takes 20 minutes at 12 km/h. He queues for 40 minutes at the post office and then cycles home at a speed of 16 km/h.

a Draw a distance–time graph representing Luigi's journey.

b At the same time as Luigi leaves home at 9.30 a.m., the postman leaves the post office. The postman walks to Luigi's house to deliver a parcel at a constant speed of 3 km/h.

   i Draw the postman's journey on the same axes.

   ii Does Luigi get home in time to sign for the parcel? Give a reason for your answer.

4 The graph shows the velocity of a cyclist during a race.

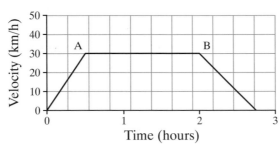

a Describe the race, giving reasons for the shape of the graph.

b How long did the race last?

c Calculate the acceleration, in km/h², of the cyclist during section:

   i OA   **ii** AB   **iii** BC

5 Two snails, Brian and Sandra, are racing along a ruler.

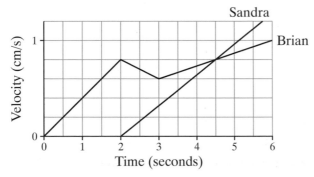

Write a commentary for the race.

You should include key velocities, times and accelerations in your commentary.

Homework 2

1   Two containers are filled with water.

Draw a graph to show how the depth (d)
of water varies with time (t).

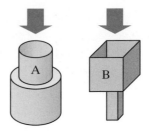

2   The surface area of a sphere varies with the radius of the sphere.

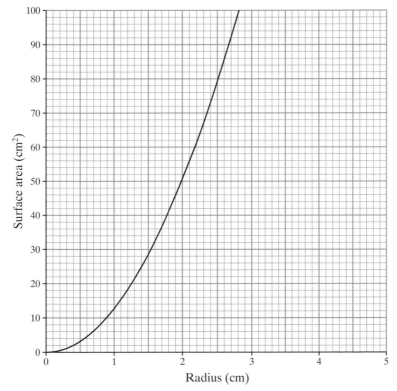

a   Use the graph to find the surface area of a sphere of radius:

   i   1 cm   ii   2 cm   iii   2.8 cm

   Give your answers to an appropriate degree of accuracy.

b   Use the graph to find the radius of a sphere with surface area:

   i   20 cm^2 ii   70 cm^2 iii   100 cm^2

   Give your answers to an appropriate degree of accuracy.

c   A factory paints ball bearings that have a diameter of 3 cm.
   One litre of paint covers 1 m^2.
   How many ball bearings can be painted from one litre?

3 The graph shows how the surface area of a square-based cuboid of volume 24 cm³ varies with the length of the cuboid.

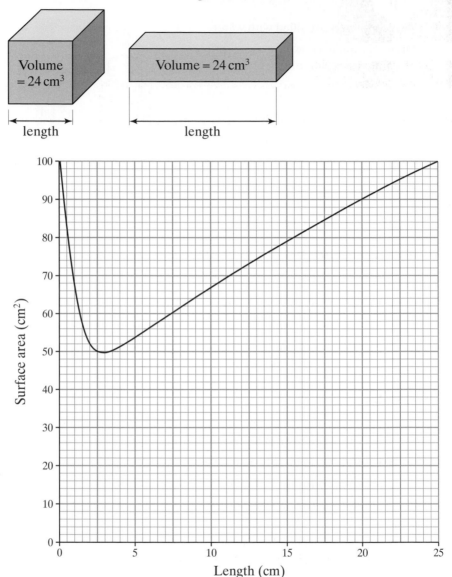

a Use the graph to find the surface area of the cuboid when the length is:

    i   1 cm   **ii**   5 cm   **iii** 20 cm

b Use the graph to find the lengths of cuboids with surface area:

    i   60 cm²   **ii**   40 cm²   **iii** 90 cm²

c Describe the shape of the cuboid whose surface area is a minimum.

6 Formulae

1 Find the value of each of these expressions if $a = 1, b = 3, c = 5$ and $d = 2$

   a $3a$ **c** $a + b + d$ **e** $4d - 3a$ **g** $b^2 - 4d$

   b $4c$ **d** $b + 4a - c$ **f** $c^2 + b - a$ **h** $b^2 - a - c$

2 If $e = 2, f = 4$ and $g = 0.5$, find the value of:

   a $4g$ **c** $f - g$ **e** $6g - 4f$ **g** efg

   b $f + 2e$ **d** $3e - g + 4f$ **f** $\dfrac{eg}{f}$ **h** $\dfrac{e^2 g}{3f}$

3 If $h = 4, l = -2, m = 3$ and $n = 1$, find the value of:

   a ln **c** $6n - 3h$ **e** $mn - l$ **g** $m^2 - h^2 - l^2$

   b $4h - 2m + l$ **d** $2h^2 + m + l$ **f** $\dfrac{h}{2} - l$ **h** $\dfrac{hmn}{l}$

4 Get Real!

By paying a deposit of £80 and then paying £50 a month for 10 months, a student can go on the school skiing trip.

   a How much would have been paid after 4 months?

   b Find a formula for the amount paid after n months.

   c Substitute $n = 10$ into your formula to find the total cost of the holiday.

5 The formula for finding the distance travelled in kilometres (d) is $d = st$, where s is the speed in km/h and t is the time in hours.
What is the distance travelled when:

   a $s = 70$ km/h and $t = 4$ h   **b** $s = 95$ km/h and $t = 1.5$ h?

6 Lucia is finding the volume of a box using the formula $V = lbh$, where V represents the volume, l represents the length, b represents the width and h represents the height of the box.
What is the volume of the box when:

   a $l = 10$ cm, $b = 8.5$ cm and $h = 3$ cm

   b $l = 6$ cm, $b = 4.5$ cm and $h = 20$ cm?

7 In geography, Nick is using the formula $C = \dfrac{5(F - 32)}{9}$

Find C if:

a $F = 50$ **b** $F = -31$

8 In maths, Emily is using the formula $A = \pi r^2$ to find the area of two circles. Take $\pi = 3.14$ and work out the area of each circle:

a $r = 14.6$ mm **b** $r = 4.5$ cm

9 In science, Oscar is using the formula $R = \dfrac{PQ}{P + Q}$

Find R if:

a $P = 5$ and $Q = 15$ **b** $P = 10$ and $Q = 2$

10 In science, Sarah is using the formula $D = \dfrac{(U + V)T}{2}$

Find D when $U = 9.5$, $V = 6.1$ and $T = 7.2$

11 Write an equation and solve it to find the unknown number in each of these questions. Let the unknown number be x.

a Think of a number. Add 12. The answer is 17.

b Think of a number. Multiply by 5. The answer is 80.

c Think of a number. Multiply by 3. Subtract 1. The answer is 5.

d Think of a number. Multiply by 9. Subtract 16. The answer is 11.

e Think of a number. Multiply by 5. Add 2. The answer is 37.

12 Write down an equation for each of these diagrams. Solve it to find the value of x.

a
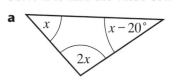

b
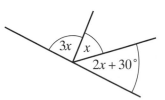

Not drawn accurately

13 The width of a rectangle is w cm. The length of the rectangle is 20 cm more than its width.

a Write an expression for the length of the rectangle.

The perimeter of the rectangle is 180 cm.

b What are the lengths of the sides?

14 A square has sides of length $(2x - 10)$ cm. The perimeter is 40 cm. What is the area of the square?

15 The length of a rectangle is double its width. The perimeter is 36 cm. What is the area of the rectangle?

Homework 2

1 Rearrange each of these formulae to make x the subject:

a $a + x = b$ **c** $e = f + x$ **e** $jx + k = l$

b $x - c = d$ **d** $g = x - h$ **f** $m + nx = p$

2 Rearrange each of these formulae to make y the subject:

a $qy - r^2 = s$ **c** $v^2 = w + xy$ **e** $\frac{1}{2}y - 3b = 4c$

b $t^2 + 4y = u^2$ **d** $zy = 4a$ **f** $d + y = e^2 - f$

3 Rearrange the formula $y = mx + c$ to make:

a c the subject   **b** m the subject.

4 Make the letter in brackets after each formula the subject:

a $GH = k$ (H) **c** $I = PRT$ (P) **e** $v = u + at$ (a)

b $E = mgh$ (h) **d** $ax + by = c$ (b) **f** $t = \dfrac{d}{s}$ (d)

5 These are all scientific formulae. Rearrange them so that x is the subject.

a $C = 2\pi x$ **c** $V = x^2 h$ **e** $e = Rx^2$

b $A = \pi x^2$ **d** $V = \frac{1}{3}x^2 h$ **f** $V = lbx$

6 Rearrange the formula $v = Ri$ to make i the subject.

7 Rearrange the formula $s = vt$ to make v the subject.

8 Rearrange the formula $v = u + at$ to make:

a u the subject   **b** a the subject   **c** t the subject.

9 Rearrange the formula $e = v + Ri$ to make:

a v the subject   **b** R the subject   **c** i the subject.

10 Rearrange the formula $y = \dfrac{2x + 3}{5}$ to make x the subject.

11 Rearrange the formula $y = \dfrac{3x - 2}{2}$ to make x the subject.

12 Rearrange the formula $a(x + 1) = b$ to make x the subject.

13 Rearrange these formulae to make x the subject.

    a $a = \dfrac{b}{x}$ **d** $u(x - v) = w$ **g** $m = xn + xp$

    b $e = \dfrac{fx}{g}$ **e** $b = \sqrt{x}$ **h** $u(x + v) = w + x$

    c $n = p(q + x)$ **f** $\sqrt{\dfrac{x}{e}} = f$

14 Rearrange $x = \dfrac{-b + \sqrt{b^2 - 4ac}}{2a}$ to make c the subject.

15 The surface area (S) of a solid cylinder is found by using the formula
$S = 2\pi r^2 + 2\pi rh$ where r represents the radius and h represents the height
of the cylinder.

    a Rearrange the formula to make h the subject.

    b Find the height of a cylinder with a radius of 4 cm and surface area of
120π cm^2.

16 Tim is using the formula $\dfrac{1}{f} = \dfrac{1}{u} + \dfrac{1}{v}$ in science.

    a Find f when $u = 4$ and $v = 12$

    b Rearrange the formula to make u the subject.

    c Find the value of u when $f = 2$ and $v = 6$

1

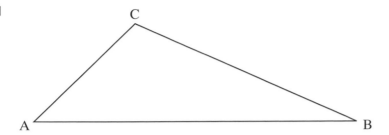

    a Measure and write down, to the nearest millimetre, the length of the sides of the triangle.

    b Measure accurately and write down the sizes of angles A, B and C.

    c Find the total of your three answers in part **b**.

2   Make an accurate drawing of parallelogram ABCD with AB = 4 cm, AD = 5 cm and BD = 4.5 cm.

    Measure and write down the length of AC on your drawing.

3   Draw accurately triangle PQR with PQ = 5.8 cm, PR = 9.4 cm and angle P = 104°.

    What is the length of QR?

4   Draw accurately an isosceles trapezium ABCD, where the base AB = 7.5 cm, the slant side AD = 4 cm and the diagonal AC = 6.5 cm.

    Measure the length of the shorter parallel side, DC.

5 **Get Real!**

    A trawler and a yacht are 4 nautical miles apart with the yacht being south-west of the trawler.

    A lighthouse is on a bearing of 245° from the trawler and on a bearing of 310° from the yacht.

    Using a scale of 1 cm to represent 1 nautical mile, draw an accurate diagram showing the positions of the trawler, the yacht and the lighthouse.

    What is the distance between the yacht and the lighthouse?

Homework 2

1 **a** Use your protractor to draw an angle of 68°.

   b Construct the angle bisector.
Check that each half angle measures 34°.

2 **a** Draw a line of length 8.4 cm.

   b Construct the perpendicular bisector of the line.
Check that each half measures 4.2 cm.

3 Get Real!

Hana would like to be selected to throw the discus in the next Olympics.

The discus arena is a V-shape with an angle of 40°.

Hana wants to keep her throws as central as possible.

Draw a diagram of the arena and construct the angle bisector so that Hana can keep a record of where her discus lands.

4 Get Real!

An artist is marking out a design based on the points of the compass on the floor of a hotel foyer.

Can you construct the same design using a ruler and compasses only?

Write out a set of instructions for someone to follow to copy your construction.

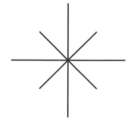

5 **a** Construct triangle PQR with PQ = 9 cm, PR = 6 cm and QR = 6 cm.

   b Construct the bisector of angle R, which cuts PQ at X.

   c Measure and write down the length of RX to the nearest millimetre.

   d Calculate the area of triangle PQR.

Homework 3

1 **a** Draw a line of length 9 cm.

   b Mark a point above the line.

   c Construct the perpendicular from the point to the line.

2 **a** Draw a line of length 8 cm.

   b Using your line as the base, construct an isosceles triangle with base angles of 45°.

3 **a** Using a ruler and compasses only, construct a square of side 5.3 cm.

   b Construct the angle bisector of one of the angles.

   If your constructions are accurate it should pass through the opposite vertex.

4 Without using a protractor, construct an angle of 135°.

5 **Get Real!**

   The diagram shows a roof truss design for a new house.

   Construct the diagram, starting with the line AB = 4 cm.

   Measure the slant height of the roof truss to the nearest millimetre.

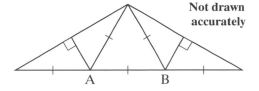

6 **Get Real!**

   A gate and a stile are 700 m apart along a straight fence. A barn is 300 m from the gate and 500 m from the stile.

   Using a ruler and compasses and an appropriate scale, construct an accurate scale drawing to find the shortest distance from the barn to the fence.

Homework 4

1 Which of these are nets of a cuboid?

a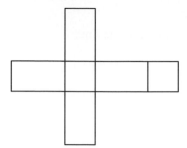

c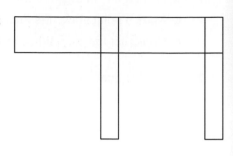

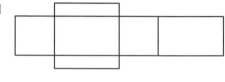

b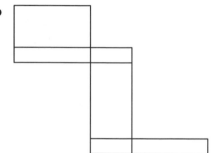

d

2 a Design nets of your own for an open cone and an open cylinder.

   b Which of these are nets of open cones, which are nets of open cylinders and which are neither?

i 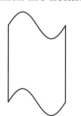   **iii** 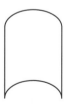   **v**

ii 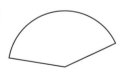   **iv**    **vi**

3 Get Real!

This is a picture of a box containing a sample tube of toothpaste.

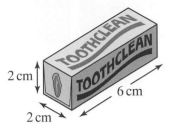

Draw an accurate net of the box using a ruler and compasses.

4 Get Real!

Olivia is designing a box to pack her homemade shortbread triangles ready to sell.

She sketches the box and adds on the measurements that she needs.

a Sketch a possible net for Olivia's shortbread box.

b Construct your net using a ruler and compasses only.

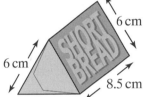

Homework 5

1 Each of the 3-D solids below is made from six centimetre cubes.

Using a centimetre grid, draw the plan of each solid and its elevation from the directions marked F and S.

a

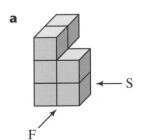

b

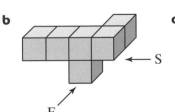

c

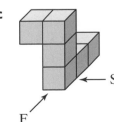

2 Ten centimetre cubes are arranged into the solid shown in the diagram.

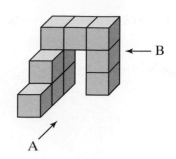

On a grid draw:

a the plan of this 3-D solid

b the front elevation as viewed from A

c the side elevation as viewed from B.

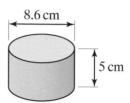

3 Get Real!

The diagram shows the dimensions of a can of tuna.

Draw an accurate plan and elevation. Leave out the hidden edges.

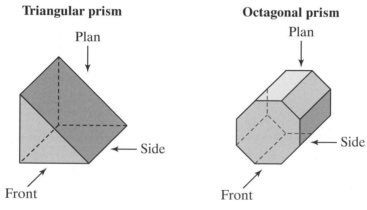

8.6 cm

5 cm

4 For each prism below, sketch the plan, front and side elevations from the directions shown by the arrows.

Triangular prism

Plan

Side

Front

Octagonal prism

Plan

Side

Front

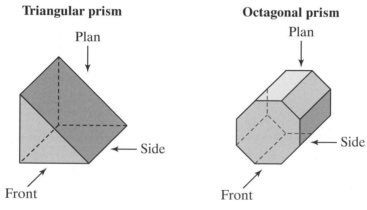

5 Each solid below is a cuboid with part removed. The dimensions are given in centimetres.

For each object, draw a full-size plan and front and side elevations on a centimetre grid.

a

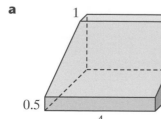

1

2

0.5

4

3

b

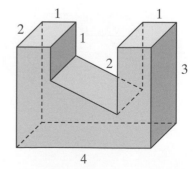

1

1

2

1

2

3

4

6 Tom has drawn a plan and a side elevation of the 3-D solid shown below.

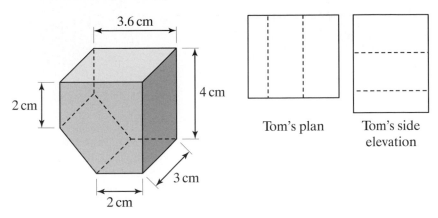

Tom's plan   Tom's side
elevation

a Describe what is wrong with Tom's diagrams.

b Draw an accurate plan and side elevation.

7 Plans and elevations of three prisms are shown below.
Sketch each object.

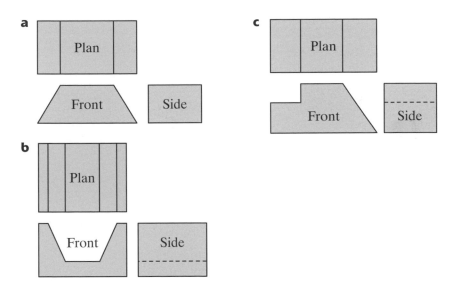

8 Get Real!

The diagram shows a toolbox.

a Sketch a plan of the toolbox.

b Sketch an elevation of the toolbox when it is viewed from the front, F.

c Sketch an elevation of the toolbox when it is viewed from the side, S.

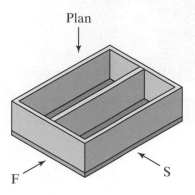

9 A prism is 8 cm long and its cross-section is an equilateral triangle with sides of length 6 cm.

Using a ruler and compasses only, construct an accurate front elevation, plan and side elevation of the prism.

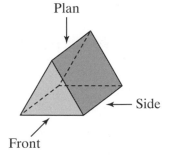

8 Probability

1   A coin is tossed 150 times. The result is 79 heads and 71 tails. Todd says, 'You should get equal numbers of heads and tails, so there's something wrong with that coin'.

    Is he correct? Give a reason for your answer.

2   If the probability of getting an unwrapped toffee in a batch is 0.005, how many unwrapped toffees would you expect to find in a batch of 1000?

3   A spinner with nine equal divisions is spun 180 times. Three divisions are labelled X, two are Y, two are Z, one is P and one is Q.

    a How many times would you expect the spinner to land on X?

    b How many times would you expect it to land on Z?

    c How many times would you expect it to land on Q?

4   The probability that an inhabitant of Random Island is left-handed is $\frac{1}{8}$ There are 4263 people living on Random Island. How many of them would you expect to be left-handed?

5   The table shows the frequency distribution after choosing a card at random from a full pack 520 times.

	Results from 520 random choices		
	Picture card (J, Q, K)	Ace	Number card
Frequency distribution	119	44	357

    a What is the relative frequency of getting an Ace?

    b Using theoretical probability, how many Aces would you expect?

    c Which one of the results in the table is the closest to the result you would expect from theoretical probability?

6 There are 3 yellow discs, 4 green discs and 5 black discs in a box. A disc is chosen at random and then put back. This is done 120 times. The table shows the frequency distribution.

	Results from choosing a disc 120 times		
	Yellow	Green	Black
Frequency distribution	33	39	48

  a What is the relative frequency of getting yellow?

  b What is the relative frequency of getting yellow or green?

  c What is the theoretical probability of getting black?

7 There are 50 counters in a bag. Some are blue, some are yellow and some are green. One counter is taken out at random and replaced. The frequency distribution table shows the results from 250 trials.

	Results from 250 trials		
	Blue	Yellow	Green
Frequency distribution	34	125	91

How many counters are there of each colour?

8 Sam has some coloured balls in a box. He asks his friends to pick a ball at random. He wants to find the probability of getting a yellow ball. After every ten attempts, he finds the relative frequency of yellow and the results are shown on the graph.

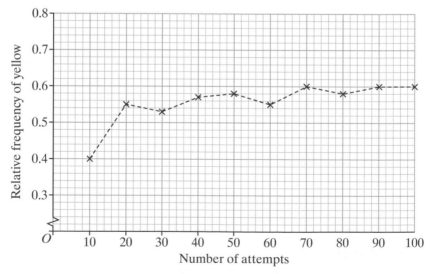

  a How many yellow balls had been picked after 10 attempts?

  b How many yellow balls had been picked after 50 attempts?

  c From the graph, estimate the probability of picking a yellow ball.

  d If there are 15 yellow balls, how many balls are in the box altogether?

Homework 2

1   **a** Draw a table to show all the possible outcomes when four coins are tossed.

   Use your table to find:

   b the probability of getting four heads

   c the probability of getting one head and three tails

   d the probability of getting two heads and two tails

   e the probability of getting at least one tail.

2   Kirsty has six tiles and two cards as shown in the diagram.

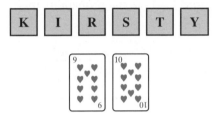

   She takes one tile and one card at random.

   a Draw a table of outcomes.

   Use your table to find the probability that Kirsty takes:

   b tile K and the 10 of hearts

   c tile S or T and the 10 of hearts

   d one of the tiles with a consonant and the 9 of hearts.

3   Two dice are thrown and the difference between their scores is recorded.

	1	2	3	4	5	6
1	0	1	2			
2						
3						
4						
5				1		
6				2		

   a Copy and complete the sample space diagram.

   Use your diagram to find:

   b the probability of getting a difference of 4

   c the probability of getting a difference of 6

   d the probability of getting a difference greater than 4.

4 Gemma has the King, Queen and Jack of hearts plus the five tiles shown in the diagram.

She picks up one tile and one card at random.

a Draw a table of outcomes.

Use your table to find the probability that Gemma picks:

b the Jack and a tile with an E

c the Jack and a tile with an M.

5 Noah picks a card at random from a full pack. Harry picks a tile at random from a bag with five Hs and three Ys. Find the probability that:

a Noah picks a heart and Harry picks a Y

b Noah picks a black card and Harry's tile is an H

c Noah does not pick a club and Harry does not pick a Y.

6 Jared has 20 marbles. Some are black, some are white and some are silver. He picks a marble at random. The probability that he will pick a black marble is 0.7 and the probability that he will pick a white marble is 0.25

a How many black marbles does he have?

b How many white marbles does he have?

c What is the probability that he will pick a silver marble?

7 A hexagonal spinner has sides numbered 2, 4, 6, 8, 10, 12. A second hexagonal spinner has sides numbered 1, 3, 5, 7, 9, 11. They are both spun and the product of their scores is recorded.

a Draw a sample space diagram.

Use your diagram to find:

b the probability of the product being an odd number

c the probability of the product being 18 or 36

d the probability of the product being a multiple of 5

e the probability of the product being at least 30.

8 Jodie throws three dice.

a How many possible outcomes are there?

b What is the probability that she gets a combined score of 18?

c What is the probability that she gets a combined score of 4?

Homework 3

1   Which of these are mutually exclusive events?

   a Throwing a two and an even number on a throw of a dice.

   b Throwing a number greater than three and an even number on a throw of a dice.

   c Picking a heart and a Queen from a pack of cards.

   d Picking the three of diamonds and the three of clubs from a pack of cards.

   e Picking a red ball and picking a blue ball from a bag containing red, white and blue balls.

2   The probability that Marie will fail her history exam is $\frac{3}{20}$
   What is the probability that she will pass the exam?

3   The probability that AQA Rovers will win their next match is 0.45
   If the probability that they will draw is 0.25, what is the probability that they will lose?

4   Dilip can get into town in three different ways – with a lift from his father, by bus or walking. The probability that Dilip will get a lift is 0.65
   The probability that he will catch the bus is 0.3
   Baz says, 'That means there is a probability of 0.32 that he will walk'.
   Explain why Baz is wrong and give the correct probability.

5   Restaurant Agaluxe has a menu with four choices.
   The probability of each dish is shown below.

Dish	Probability
Steak	0.57
Chicken	0.33
Monkfish	0.15
Vegetarian	0.05

   a How can you tell that the probabilities are incorrect?

   b The first probability is incorrect.
      What should it be?

6   Vicky goes out to buy some new jeans.
   The probability that she buys them from Jeans'R'Us is $\frac{1}{4}$
   The probability that she buys them from the Superstore is $\frac{5}{12}$
   What is the probability that she buys them from one of these stores?

7 The probability that Jared will go to Leeds University is 0.45
The probability that he will go to Sheffield University is 0.25 and
the probability that he will go to Manchester University is 0.2

   a What is the probability that Jared will go to either Leeds or
Manchester?

   b What is the probability that he does not go to Manchester or
Sheffield?

   c What is the probability that he does not go to any of these three
universities?

8 Joanne has a box of fruit containing 8 apples, 11 pears, 6 oranges and
5 kiwi fruits. She takes a fruit from the box at random.

   a What is the probability that she takes a kiwi fruit?

   b What is the probability that she does *not* take an orange?

   c What is the probability that she takes either an apple or an orange?

9 Sam has a bag of 60 coloured marbles. Some are blue, some are green,
some are white and the rest are multi-coloured. Sam picks a marble at
random from the bag.
The probability that he will pick a blue marble is 0.15

   a How many blue marbles are in the bag?

   The probability that he will *not* pick a green marble is 0.75

   b How many green marbles are in the bag?

   Sam is twice as likely to pick a multi-coloured marble as a white one.

   c How many white marbles are in the bag?

10 Mr Mathsman has a collection of 250 books. Some are cookery books,
some are gardening books, 74 of them are travel books and the rest are
books about chess. He has three times as many gardening books as
cookery books. If one book is picked at random, the probability that it
will be a book about chess is 0.32

   a How many chess books does he have?

   b How many gardening books does he have?

Homework 4 🖩

Apart from question 8 this is a non-calculator exercise.

You may assume that the events described in this exercise are independent.

1 Karl takes a ball at random from a bag containing 2 red balls and 3 black ones. Errol throws a dice. What is the probability that:

 a Karl takes a red ball and Errol throws a six?

 b Karl takes a black ball and Errol throws an odd number?

 c Karl does not take a black ball and Errol does not throw a six?

2 There is a probability of 0.9 that Tanya will come top in the science exam and a probability of 0.4 that she will come top in geography. What is the probability that she comes top in both subjects?

3 The probability that Jan eats an egg for breakfast is $\frac{4}{5}$
 The probability that she cycles to school is $\frac{5}{6}$
 What is the probability that Jan eats an egg and cycles to school?

4 The probability that Dave gets into the rugby team is $\frac{3}{8}$
 The probability that he gets an A grade in his maths exam is $\frac{5}{6}$
 What is the probability that Dave gets into the rugby team but does not get an A grade in his maths exam?

5 Two dice are thrown together.
 What is the probability that both show an even number?

6 Five coins are tossed simultaneously.
 What is the probability that they all show heads?

7 If n coins are tossed simultaneously what is the probability that they all show heads?

🖩 **8** The probability that Jake will go to the gym in the evening is 0.6
 What is the probability that he goes to the gym on Monday, Tuesday and Thursday evenings but not Wednesday evening?

9 Pippa, Roz and Sally are all taking their ballet exam next week.
The probability that Pippa will pass is $\frac{2}{3}$
The probability that Roz will pass is $\frac{3}{4}$
The probability that Sally will pass is $\frac{5}{6}$

    a Tara says that the probability that they will all pass is $\frac{2}{3} + \frac{3}{4} + \frac{5}{6} = \frac{27}{12}$

        i Why should Tara know that her answer must be wrong?

        ii What is the correct answer?

    b What is the probability that Pippa and Sally will pass but Roz will fail?

    c What is the probability that they will all fail?

Homework 5

1 The probability that Aisha will pass her piano exam is 0.9 and the probability that Sean will pass his clarinet exam is 0.7

    a Copy and complete the tree diagram.

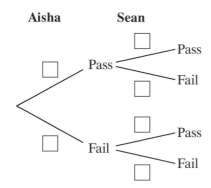

    Use it to find:

    b the probability that both of them pass their exams

    c the probability that Aisha fails and Sean passes

    d the probability that at least one of them fails.

2 A box contains 7 blue pens and 9 black pens. One pen is taken from the box and then replaced. A second pen is then taken from the box.

    a Draw a tree diagram.

    Use it to find:

    b the probability that both pens are blue

    c the probability that the first pen is blue and the second pen is black

    d the probability that one pen is blue and one is black.

3 Do question **2** again on the basis that the first pen is *not* replaced.

4 There are six brown eggs and six white eggs in a box. Two eggs are taken at random. Find the probability that:

 a they are both white

 b one egg is brown and one is white.

5 A box contains 15 milk chocolates and 10 dark chocolates. Imogen takes a chocolate at random and eats it. If it is milk chocolate, she will take another one. If it is dark chocolate, she will stop. Find the probability that Imogen takes:

 a two milk chocolates

 b two dark chocolates.

6 The probability that Tara's Mum will pack her lunch box overnight is 0.8
If Tara finds her lunchbox already packed, the probability that she will get to school on time is 0.9, but if Tara has to pack her own lunch box, the probability that she will be on time is 0.4
Find the probability that Tara is late for school.

7 The probability that Frazer will go swimming on Saturday is $\frac{4}{5}$
If he goes swimming, the probability that he will go for a run is $\frac{1}{4}$
If he does not go swimming, the probability that he goes for a run is $\frac{5}{6}$
Find the probability that Frazer does not go for a run on Saturday.

8 Paula is going to France for a holiday. She can travel by ferry, plane or rail. The probability that she goes by ferry is 0.2 and the probability that she goes by plane is 0.45

 If Paula goes by ferry, the probability that she stays a night in Paris is 0.8, but if she goes by plane, the probability that she stays a night in Paris is 0.3

 If she goes by rail, the probability that she stays a night in Paris is 0.5

 Find the probability that:

 a Paula goes by ferry but does not stay in Paris

 b she goes by rail and stays in Paris

 c she stays in Paris.

9 Joe has 4 plain white socks, 8 grey socks and 12 patterned socks in his drawer. He takes out one sock and then another.

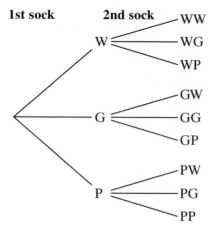

1st sock **2nd sock**
WW
WG
WP
GW
GG
GP
PW
PG
PP

a Copy and complete the tree diagram.

Find:

b the probability that Joe has picked two patterned socks

c the probability that he has picked one white and one grey sock

d the probability that he has picked a pair of matching socks.

10 Box A contains one black ball, one pink ball, one yellow ball, one blue ball and one green ball. Box B contains one black ball, one pink ball, one yellow ball, one blue ball and two white balls. One ball is picked at random from box A and placed in box B. A ball is then picked at random from box B.

Calculate the probability that:

a the ball picked from box A is blue and the ball picked from box B is also blue

b the ball picked from box A and the ball picked from box B are the same colour.

9 Vectors

Homework 1

1 Write the vectors shown in the diagram as column vectors.

2 Use your answer for question **1** to calculate:

   a $\mathbf{a} + \mathbf{b}$ **d** $\mathbf{c} - \mathbf{a} + \mathbf{d}$

   b $\mathbf{b} + \mathbf{c}$ **e** $\mathbf{h} - \mathbf{c} + \mathbf{d}$

   c $\mathbf{g} - \mathbf{d}$

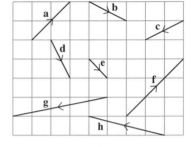

3 $\mathbf{p}$ is the vector $\begin{pmatrix} 2 \\ -1 \end{pmatrix}$, $\mathbf{r}$ is the vector $\begin{pmatrix} -3 \\ -2 \end{pmatrix}$.

If $\mathbf{p} + \mathbf{q} = \mathbf{r}$, write $\mathbf{q}$ as a column vector.

4 Darren is taking a penalty. He tries to place the ball inside the right-hand post, by shooting with the vector $\begin{pmatrix} 3 \\ 14 \end{pmatrix}$. However, the wind is blowing with a strength and direction $\begin{pmatrix} 2 \\ -1 \end{pmatrix}$.

Find where the ball goes, and decide whether he scores.

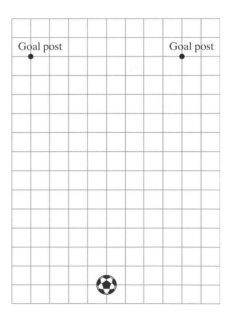

5 Simplify these vectors.

a $4\begin{pmatrix} 2 \\ 3 \end{pmatrix}$
b $3\begin{pmatrix} 3 \\ -2 \end{pmatrix}$
c $3\begin{pmatrix} -2.5 \\ 4 \end{pmatrix}$
d $-3\begin{pmatrix} 3 \\ 2.5 \end{pmatrix}$

6 If $\mathbf{a} = \begin{pmatrix} -2 \\ 3 \end{pmatrix}$, $\mathbf{b} = \begin{pmatrix} -1 \\ -2 \end{pmatrix}$ and $\mathbf{c} = \begin{pmatrix} 3 \\ -4 \end{pmatrix}$, simplify:

a $3\mathbf{a}$ **c** $\mathbf{a} + \mathbf{b} + \mathbf{c}$ **e** $3\mathbf{a} - \mathbf{c}$ **g** $3\mathbf{a} - 2\mathbf{b}$

b $2\mathbf{b}$ **d** $2\mathbf{a} - 2\mathbf{b}$ **f** $\mathbf{a} - \mathbf{b} + \mathbf{c}$ **h** $2\mathbf{a} - 3\mathbf{b} + \mathbf{c}$

7 If $\begin{pmatrix} a \\ a \end{pmatrix} + \begin{pmatrix} a \\ b \end{pmatrix} + \begin{pmatrix} b \\ -a \end{pmatrix} + \begin{pmatrix} -b \\ a \end{pmatrix} = \begin{pmatrix} 6 \\ 8 \end{pmatrix}$, find the value of a and b.

Homework 2

1 **a** Draw these vectors.

$$\mathbf{p} = \begin{pmatrix} 2 \\ -3 \end{pmatrix} \quad \mathbf{q} = \begin{pmatrix} 3 \\ -2 \end{pmatrix} \quad \mathbf{r} = \begin{pmatrix} 4 \\ 6 \end{pmatrix} \quad \mathbf{s} = \begin{pmatrix} 6 \\ -4 \end{pmatrix} \quad \mathbf{t} = \begin{pmatrix} 3 \\ 2 \end{pmatrix}$$

b From the vectors **p**, **q**, **r**, **s** and **t**, write down a pair of vectors that are:

i parallel

ii perpendicular.

2 ABC is a triangle where $\overrightarrow{AB} = \begin{pmatrix} 5 \\ 7 \end{pmatrix}$ and $\overrightarrow{AC} = \begin{pmatrix} 1 \\ -3 \end{pmatrix}$

Find:

a $\overrightarrow{BC}$

b $\overrightarrow{BD}$, where D is the midpoint of BC

c $\overrightarrow{AD}$

3   Express **p**, **q**, **r** and **s** in terms of **a** and **b**.

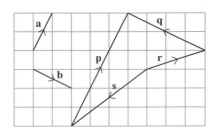

4   If $\begin{pmatrix} 4 \\ -2 \end{pmatrix}$ is parallel to $\begin{pmatrix} x \\ 6 \end{pmatrix}$, calculate x.

5   **p** = 3**a** – **b**, **q** = 2**a** – 2**b** and **r** = **a** + 3**b**

Show that 4**p** – 5**q** is parallel to **r**.

6 $\overrightarrow{OA} = \begin{pmatrix} 3 \\ 1 \end{pmatrix}$, $\overrightarrow{OB} = \begin{pmatrix} 1 \\ -2 \end{pmatrix}$ and $\overrightarrow{OC} = \begin{pmatrix} -4 \\ 3 \end{pmatrix}$.

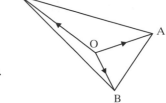

M, N are the midpoints of AC and AB respectively.

Write, as column vectors:

a $\overrightarrow{AB}$ **c** $\overrightarrow{AC}$ **e** $\overrightarrow{OM}$ **g** $\overrightarrow{ON}$

b $\overrightarrow{BC}$ **d** $\overrightarrow{BM}$ **f** $\overrightarrow{AN}$ **h** $\overrightarrow{MN}$

Homework 3

1   OABC is a rectangle. $\overrightarrow{OA}$ = **a** and $\overrightarrow{OC}$ = **b**.

D is the point on OA such that OD = 3DA, and E is the point on AB such that AE = 2EB.

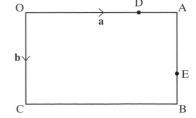

Find, in terms of **a** and **b**:

a $\overrightarrow{OD}$

b $\overrightarrow{CD}$

c $\overrightarrow{BE}$

d $\overrightarrow{CE}$

e $\overrightarrow{OM}$, where M is the midpoint of DE.

2 ABCD is a parallelogram.

CBE is a straight line, such that BE = 2CB.

F is the midpoint of AB.

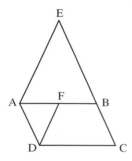

If $\overrightarrow{DA}$ = **p** and $\overrightarrow{AB}$ = **q**, write, in terms of **p** and **q**:

a $\overrightarrow{DF}$

b $\overrightarrow{AE}$

c What can you say about DF and AE?

3 ABCD is a parallelogram.

P is the point on AB such that AP = 2PB.

Q is the point on DC such that DQ = 2QC.

R is the point on QP such that RP = 2QR.

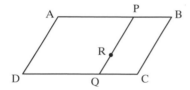

If $\overrightarrow{CB}$ = **p** and $\overrightarrow{QC}$ = **q**, find, in terms of **p** and **q**:

a $\overrightarrow{DC}$

b $\overrightarrow{AD}$

c $\overrightarrow{AC}$

d $\overrightarrow{AP}$

e $\overrightarrow{PR}$

f $\overrightarrow{AR}$

g Hence show that A, R and C are collinear.

10 Graphs of linear functions

Homework 1

1 Draw an *x*-axis and a *y*-axis, each going from −6 to 6.

  a On the axes, draw the lines *x* = 5 and *x* = −3

  b On the axes, draw the lines *y* = −5 and *y* = 4

  c Where do *x* = 5 and *y* = 4 meet?

  d Where do *y* = 4 and *x* = −3 meet?

2 **a** Copy and complete this table of values for *y* = 3*x*

x	−3	−2	−1	0	1	2	3
y		−6			3		

  b Draw an *x*-axis labelled from −3 to 3 and a *y*-axis labelled from −9 to 9. Plot the points in the table and join them with a straight line.

  c What are the coordinates of the point on the line where *y* = 3?

  d What are the coordinates of the point on the line where *x* = −2?

3 **a** Copy and complete this table of values for *y* = *x* − 2

x	−3	−2	−1	0	1	2	3
y		−4			−1		

  b Draw an *x*-axis labelled from −3 to 3 and a *y*-axis labelled from −5 to 5. Plot the points in the table and join them with a straight line.

  c Where does the line *y* = *x* − 2 cut:

    i the *x*-axis   **ii** the *y*-axis   **iii** the line *x* = 2   **iv** the line *y* = −1?

  d What are the coordinates of the point on the line where *x* = −2?

4 **a** Copy and complete this table of x-values and y-values for the graph
with equation $y = 3x - 3$

x	−2	−1	0	1	2	3
y	−9			0		

  b The x-coordinates go up one unit each time. How many units do the
y-coordinates go up each time?

  c Plot the points on graph paper and join them to make the straight-line graph
$y = 3x - 3$

  d Find the y-coordinate of a point on the graph with an x-coordinate of 1.5

  e Find the x-coordinate of a point on the graph with a y-coordinate of 2.

5 **a** For the graph of $x + y = 5$, complete these coordinate pairs:

$(-4, \underline{\quad})$ $(0, \underline{\quad})$ $(4, \underline{\quad})$

This means that the x-coordinates
and the y-coordinates add up to 5

  b Draw an x-axis and a y-axis, each going from −10 to 10, then plot the three
points and join them with a straight line.

  c Write down the coordinates of the points where the graph $x + y = 5$ crosses
the x-axis and the y-axis.

  d Use the equation of the graph to work out the coordinates of the points
where the graph crosses the x-axis and the y-axis.
Show how you found your answers.

6 Draw an x-axis and a y-axis, each going from −10 to 10.

  a On the axes, draw these lines:

  i $y = 4x$ **ii** $y = 2x - 1$

  b i Where do the lines $y = 4x$ and $y = 2x - 1$ cut the x-axis and the y-axis?

  ii How can you find this from the equations?

  c Where do the lines $y = 4x$ and $y = 2x - 1$ meet?

7 Which of these points lie on the graph of $y = -\frac{1}{2}x$?

$(2, -1)$, $(2, 0)$, $(0, 0)$, $(-2, -1)$, $(1, -2)$, $(2, -4)$, $(0.6, 0.3)$

Show how you found your answers.

8 **a** Copy and complete these coordinate points for the line $y = 2x - 2$

(−2, __), (0, __), (__, 0)

b Draw an x-axis and a y-axis, each going from −6 to 6, then plot the points and join them with a straight line.

c Where does the line $y = 2x - 2$ cut:

i the x-axis   **ii** the y-axis   **iii** the line $x = 2$ **iv** the line $y = 2$?

9 **a** Find at least three points on the line $y = 2x - 3$

b Draw an x-axis and a y-axis, each going from −10 to 10.
Plot the points and join them with a straight line.

c On the same diagram, draw the graph $y = 2x - 1$

d What is the same and what is different about the two graphs?

10 **a** Draw an x-axis and a y-axis, each going from −6 to 6.
Draw the graphs $y = -x + 4$ and $y = -x - 4$

b What is the same and what is different about the two graphs?

c Explain how you can use the equations to tell you what is the same and what is different about the graphs.

11 The diagram shows several straight-line graphs.

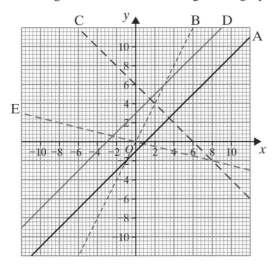

a Write down the letters of two graphs that are parallel.

b Write down the letters of two graphs that are perpendicular.

c Write down the letters of the graphs that go through the origin.

d Write down the letter of the steepest line.

e **i** Write down the coordinates of three points on graph A.

ii What is the equation of graph A?

Homework 2 🔲

1 Which of these equations represent straight-line graphs?

    a $x + y = 5$ **b** $xy = 5$ **c** $y = \dfrac{x}{5}$ **d** $y = \dfrac{5}{x}$ **e** $y = x^2 + 5$

2 Rearrange each of these equations into the form $y = mx + c$

    a $x = 2y$ **b** $x + 2y = 0$ **c** $x + 2y = 3$ **d** $x - 2y = 0$

3 **a** Copy and complete the table of x-values and y-values for the graphs with equations $y = 3x$ and $y = 3x - 4$

x	-2	-1	0	1	2	3
$y = 3x$			0			
$y = 3x - 4$			-4			

    b Draw both graphs on the same diagram.

    c Use your diagram to find:

       i the gradient of each graph

      ii the point where each graph cuts the y-axis.

4 Repeat question **3** for the graphs $y = \frac{1}{2}x$ and $y = \frac{1}{2}x + 3$

5 Find the gradients of the straight-line graphs with these equations.

    a $y = 3x$ **e** $y = 3x + 5$ **i** $y = \frac{1}{5}x$ **m** $y = -0.3x$

    b $y = \frac{1}{4}x$ **f** $y = \frac{1}{4}x + 5$ **j** $y = -\frac{1}{5}x$ **n** $y = 0.3x - 0.2$

    c $y = -3x$ **g** $y = -3x + 5$ **k** $y = \frac{1}{4}x - \frac{4}{5}$ **o** $y = 0.3x$

    d $y = -\frac{1}{4}x$ **h** $y = -\frac{1}{4}x + 5$ **l** $y = -\frac{1}{5}x - \frac{4}{5}$ **p** $y = -0.3x - 0.2$

6 Find the coordinates of the points where the lines in question **5** meet the y-axis.

7 **a** Write down the gradients and *y*-intercepts of the graphs in this diagram.

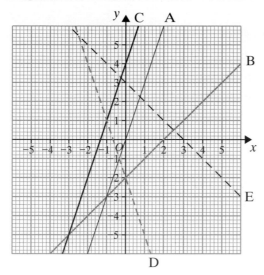

The *y*-intercept is the *y*-coordinate of the point where the line cuts the *y*-axis.

b Write down the equation of each graph.

8 **a** Rearrange each of these equations into the form $y = mx + c$

  i $y + x = 15$ **ii** $y - x = 15$ **iii** $y + x = -15$ **iv** $y - x = -15$

  b Find the gradients of the graphs represented by these equations.

9 Rearrange each of these equations into the form $y = mx + c$ and find the gradients and *y*-intercepts of the corresponding straight-line graphs.

  a $3y + x = 9$ **c** $3y - x = -9$ **e** $3x + 5y = 15$ **g** $-3x + 5y = -15$

  b $y - 3x = 9$ **d** $y + 3x = -9$ **f** $3x - 5y = 15$ **h** $3x + 5y + 15 = 0$

10 **Get Real!**

The equation $y = \frac{8}{5}x$ gives a straight-line graph that converts distances in miles (*x*) to distances in kilometres (*y*).

  a What is the gradient of the graph?

  b What is its *y*-intercept?

  c Interpret the meanings of these numbers.

  d Copy and complete this table of values for the graph.

Distance (miles)	0	10	
Distance (kilometres)			40

  e Draw the graph, choosing scales so that the three points in the table fit on the diagram.

  f Use the graph to convert 44 miles to kilometres.

  g What graph would convert distances in kilometres to distances in miles?

Homework 3 ▨

1 Find the equations of these lines:

a Gradient of 2;   goes through $(0, 0)$

b Gradient of 2;   goes through $(0, 5)$

c Gradient of 2;   goes through $(0, -2)$

d Gradient of -3;   goes through $(0, 0)$

e Gradient of -3;   goes through $(0, -3)$

f Gradient of -3;   goes through $(0, 4)$

g Gradient of -1;   goes through $(0, 7)$

h Gradient of $\frac{1}{4}$;   goes through $(0, -4)$

i Gradient of -1.4;   goes through $(0, 0.7)$

j Gradient of -2;   goes through $(0, 0)$

2 Find the gradient of the straight line that goes through each of the following pairs of points.

a $(0, 0)$ and $(4, 12)$

b $(0, 0)$ and $(12, 4)$

c $(0, 5)$ and $(5, 0)$

d $(3, 5)$ and $(7, 7)$

e $(4, 5)$ and $(-2, -4)$

3 Find the equations of the lines in question **2**.

4 The diagram shows some parallel and perpendicular straight-line graphs.

   a Use the diagram to identify pairs of lines that are parallel and pairs of lines that are perpendicular.

   b Find the gradient of each line and check that the gradients of the parallel lines are the same and that the gradients of perpendicular lines have a product of −1.

   c Find the equation of each line.

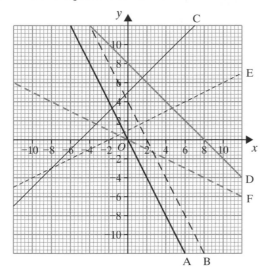

5 These equations represent four sets of parallel lines.
Sort the equations into the four sets.

$y = 4x$ $y = 7 - 4x$ $y = \frac{1}{4}x$ $x + 5 = \frac{1}{4}y$

$y - 4x = 5$ $4x + y - 5 = 0$ $x + \frac{1}{4}y = 1$ $y = 4x + 3$

$4y - x = 2$ $4y = x + 3$ $y = -4x$ $y = -\frac{1}{4}x$

$y = \frac{1}{4}(x + 2)$ $y = \frac{1}{4}(5 - x)$ $y = 3 - \frac{1}{4}x$ $4y + x = 2$

6 Here are the equations of three straight-line graphs. For each pair of graphs, decide whether they are parallel, perpendicular or neither.

   $4x + 3y = 10$, $4x - 3y = 5$, $3y = 12 - 4x$

Give a reason for your answer.

7 On graph paper:

  a draw the line $5x + 4y = 20$

  b i draw a line parallel to $5x + 4y = 20$ that passes through the point $(4, 5)$

    ii find the equation of this line

  c i draw a line perpendicular to $5x + 4y = 20$ that passes through the point $(4, 5)$

    ii find the equation of this line.

8 Find the equation of the perpendicular bisector of the line segment joining each pair of points.

  a $(0, 0)$ and $(4, 8)$ **d** $(-5, 5)$ and $(5, -5)$ **g** $(-7, -7)$ and $(7, 7)$

  b $(0, 0)$ and $(4, 4)$ **e** $(-8, 8)$ and $(8, -8)$

  c $(8, 0)$ and $(0, 8)$ **f** $(-5, -5)$ and $(5, 5)$

9 A triangle has its vertices at the points $(-6, 4)$, $(-2, -2)$ and $(2, 0)$. Find whether the triangle is right-angled or not, showing your working.

10 The equation $x + y = a$, where a is a constant, is the equation of a straight-line graph.

Complete the table to show whether the equations represent graphs parallel to $x + y = a$, perpendicular to $x + y = a$ or neither.

Equation of line	Parallel to $x+y=a$	Perpendicular to $x+y=a$	Neither parallel nor perpendicular
$y = x + 5$			
$y = -x$			
$y = -x + 2a$			
$y + 2x = a$			
$x - y = a$			
$x + 2y = a$			
$x = 5 + y$			
$x = -y$			

Find any equations that represent the same graph.

11 Similarity and congruence

Homework 1

Apart from question 1, this is a calculator exercise.

 1 **i** Explain why the following pairs of shapes are **not** similar.

  ii Sketch the second shape of each pair and alter one measurement so that the pair are similar.

a

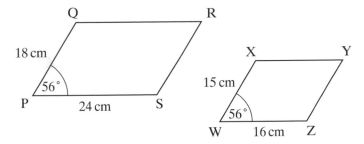

b

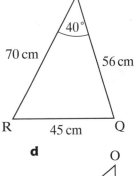

c

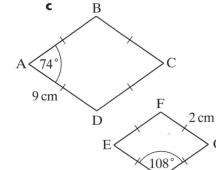

d

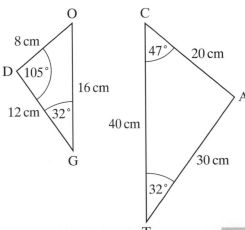

2 **a** Name two similar triangles in the diagram.

b Name the angle equal to angle AMS.

c If MS = 3.6 cm and AS = 2.4 cm, calculate ST.

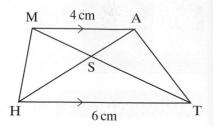

3 Find the lengths of the marked sides in these diagrams.

a

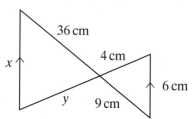

c

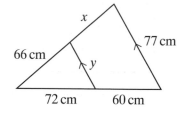

b

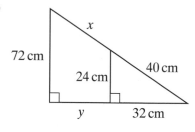

d

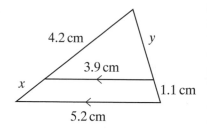

<u>**4**</u> **Get Real!**

Callum has designed a logo for members of the Three Peaks Climbing Group.

AG, BF and CE are all parallel.

AB = 2 cm, BC = 3 cm, CD = 4 cm and ED = 6 cm.

a Calculate FE.

b Calculate GF.

c Triangles BDF and ADG are similar.

What is the ratio of their corresponding sides?

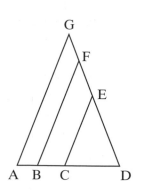

Homework 2

Apart from question 3, this is a calculator exercise.

1 Copy and complete the table.

Linear scale factor	Linear ratio	Area scale factor	Area ratio	Volume scale factor	Volume ratio
2	1:2	4	1:4	8	1:8
3			1:9		
				125	
			1:100		
		625			
$\frac{1}{2}$	2:1	$\frac{1}{4}$			
			16:1	$\frac{1}{64}$	64:1
$\frac{1}{3}$		$\frac{1}{9}$			27:1
			25:1		
		$\frac{4}{9}$			
	3:5				
$\frac{9}{4}$					

2 The hypotenuses of two similar right-angled triangles are in the ratio 4:7.
The area of the larger triangle is 78.4 cm².
What is the area of the smaller triangle?

3 Two hexagonal prisms are similar.

The smaller prism has dimensions that are half those of the larger prism.

The volume of the smaller prism is 50 cm³.

Calculate the volume of the larger prism.

4 Get Real!

A catering company stores cooking oil in two similar containers.

The larger container has a base area of 680 cm² and holds 53 litres of oil.

If the smaller container has a base area of 306 cm², how many litres of oil does it hold?

Give your answer to the nearest whole number.

Homework 3

1 Get Real!

A farmer needs to know the
distance across a marsh, from A
to B, in one of his fields.

He makes a diagram and
measures the distances shown.

By first proving that the triangles
ABX and ZYX are congruent,
find AB.

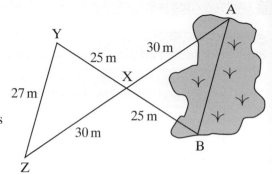

2 Prove that there is a pair of congruent triangles in each of these
diagrams.

Some equal sides and angles are marked.

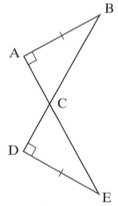

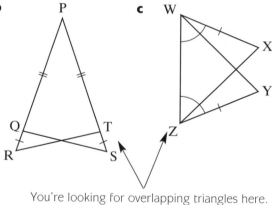

You're looking for overlapping triangles here.

3 In the rhombus ABCD prove that triangle ABD is congruent to
triangle CBD.

4 Triangles ABC and BDF are two
overlapping equilateral triangles.

Prove triangle BCF ≡ triangle BAD.

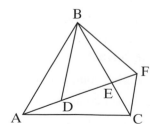

12 Pythagoras' theorem

Homework 1

1 Draw a sketch and find the length of the missing side of each triangle.

Give your answer to an appropriate degree of accuracy.

  a AB = 20 cm, AC = 12 cm, angle ACB = 90°

  b XY = 20 cm, YZ = 29 cm, angle YXZ = 90°

  c DE = 20 cm, EF = 15 cm, angle DEF = 90°

  d PR = 20 cm, RQ = 48 cm, angle PRQ = 90°

  e JK = 20 cm, KL = 20.5 cm, angle KJL = 90°

  f VW = VX = 20 cm, angle WVX = 90°

  g ST = 20 cm, US = UT, angle SUT = 90°

 2 Get Real!

A gate is 2 metres wide.

The horizontal bars are 17 cm apart and each of the six bars is 2.5 cm thick.

Find the length of the diagonal (ignore its width).

Leave your answer as a square root.

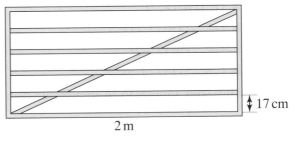

2 m

17 cm

3 Get Real!

The country of Sylvania has declared itself a no fly zone and will no longer allow aircraft to fly through its airspace.

Captain Walkington has had to submit a new flight plan for his journey.

He will fly 300 km south, then turn and fly 250 km east.

How many extra kilometres will he fly using this flight plan instead of flying direct?

13 Quadratic graphs

Homework 1

1 The graph shows the function $y = -x^2$ for values of x from -4 to 4.

Graph of $y = -x^2$

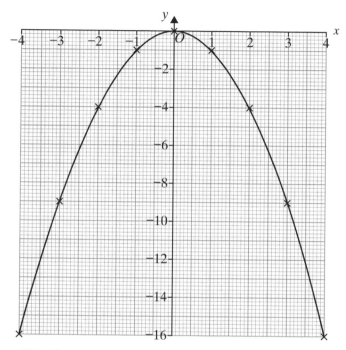

a Use the graph to estimate the value of:

   i -3.6^2 **ii** $-(-2.4)^2$

b Use the graph to find the values of x when:

   i $y = -11$ **ii** $y = -7.8$

2   **a** Copy and complete this table for $y = x^2 - 3$

x	-3	-2	-1	0	1	2	3
y		1	-2				6

   b Draw the graph of $y = x^2 - 3$ for values of x from -3 to 3.

   c Use your graph to find the value of y when:

   i $x = 1.6$ **ii** $x = -2.3$

   d Use your graph to find the values of x when:

   i $y = 3.8$ **ii** $y = -1.4$

 3   **a** Draw the graph of $y = 4x^2$

   b Compare your graph with that of $y = -x^2$
   What are the similarities and differences?

 4   **a** Copy and complete the table for $y = 3x^2 + 1$

x	-4	-3	-2	-1	0	1	2	3	4
$3x^2$	48	27				3	12		
$y = 3x^2 + 1$	49	28				4	13		

   b Draw the graph of $y = 3x^2 + 1$ for values of x from -4 to 4.

   c Give the y-coordinate of the point on the curve with an x-coordinate of:

   i   3.5   **ii** -2.4

   d Give the x-coordinate of the points on the curve with a y-coordinate of:

   i   25   **ii**   7

 5   **a** Draw the graph of $y = 2x^2 - 5$ for values of x from -4 to 4.

   b Use your graph to find the value of y when:

   i $x = 1.3$ **ii** $x = -3.2$

   c Use your graph to find the values of x when:

   i $y = 20$ **ii** $y = 9$

   d Write the coordinates of the lowest point on the curve.

 6   **a** Draw the graph of $y = 14 - x^2$ for values of x from -4 to 4.

   b i   Give the x-coordinates of the points where the curve crosses the x-axis.

   ii   Explain why the answers to part **i** are the square roots of 14.

7 **a** Copy and complete the table below, then use it to draw the graph of $y = (x + 1)(4 - x)$

x	−2	−1	0	1	1.5	2	3	4	5
$x + 1$	−1				2.5		4		6
$4 - x$	6				2.5		1		−1
$y = (x + 1)(4 - x)$	−6				6.25		4		−6

b Write down the coordinates of the points where the curve crosses the x-axis.

 8 **a** Draw a table and a graph for $y = x(x + 2)$ for values of x from −4 to 2.

b Write down the coordinates of the points where the curve crosses the x-axis.

 9 **Get Real!**

When a stone is dropped from a high bridge over a river, the height of the stone above the water, h metres, after t seconds is given by the formula $h = 76 - 5t^2$

a Draw the graph of $h = 76 - 5t^2$ for $0 \leqslant t \leqslant 4$

b Use your graph to find:

i the height of the stone after 1.5 seconds

ii the time when the stone is 20 metres above the water

iii the time when the stone hits the water.

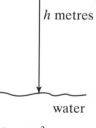

h metres

water

10 The graph shows the points that Cliff has plotted for his graph of $y = 2x^2$

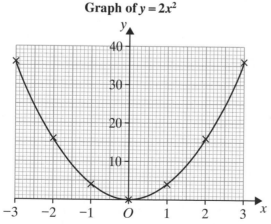

Graph of $y = 2x^2$

a Is Cliff correct?

b Give a reason for your answer.

11 The graphs of three quadratic functions are shown in the sketch.

The functions are

$$y = x^2 + 5 \qquad y = 5x^2 \qquad y = -5x^2$$

Choose the function that represents each curve.

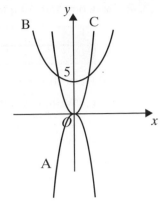

 12 **a** On separate axes draw the graphs of:

i $y = (2+x)(4+x)$ **iii** $y = (2+x)(4-x)$

ii $y = (2-x)(4-x)$ **iv** $y = (2-x)(4+x)$

b What do you notice about your graphs?

 13 **a** Give four possible equations for graph A.

b Give four possible equations for graph B.

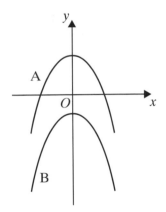

Homework 2

 1 **a** Copy and complete this table for $y = x^2 + 2x$

x	−4	−3	−2	−1	0	1	2
y	8	3		−1		3	

b Draw the graph of $y = x^2 + 2x$ for values of x from −4 to 2.

c Use your graph to solve the equation $x^2 + 2x = 0$

d **i** Draw the line $y = 7$ on your graph.

ii Find the x-coordinates of the points where the line $y = 7$ crosses the curve $y = x^2 + 2x$

iii Write down the quadratic equation whose solutions are the answers to part **ii**.

2   **a** Copy and complete this table for $y = x^2 - 2x + 1$

x	-2	-1	0	1	2	3	4
y	9				1	4	

b Draw the graph of $y = x^2 - 2x + 1$ for values of x from -2 to 4.

c i   Write down the x-coordinate of the point where the curve meets the x-axis.

ii Write down the quadratic equation whose solution is the answer to part **i**.

d i   Write the x-coordinates of the points where the line $y = 5$ crosses the graph of $y = x^2 - 2x + 1$

ii Write a quadratic equation whose solution is the answer to part **i**.

 3   **a** Copy and complete this table for $y = 10 - x - x^2$

x	-4	-3	-2	-1	-0.5	0	1	2	3
y		4			10.25			4	

b Draw the graph of $y = 10 - x - x^2$ for values of x from -4 to 3.

c i   Find the x-coordinates of the points where the graph crosses the line $y = 0$

ii Write down the quadratic equation whose solutions are the answers to part **i**.

d Use your graph to find the solutions of:

   i $4 - x - x^2 = 0$ **ii** $x^2 + x = 3$ **iii** $x^2 + x - 7 = 0$

 4   **a** Draw the graph of $y = 2x^2 + x - 7$ for values of x from -4 to 4.

b i   Write the x-values where the curve crosses the x-axis.

ii Write the quadratic equation whose solutions are the answers to part **i**.

c i   Write the x-values where the curve meets the line $y = 15$

ii Write a quadratic equation whose solutions are the answers to part **i**.

5 **Get Real!**

When a car does an emergency stop, the distance it travels in coming to a halt is related to its speed by the formula:

$d = 0.015v^2 + 0.3v$ where d is the distance in metres and v is the speed in miles per hour.

a Draw a graph of d against v for values of v from 0 to 70.

b Use your graph to estimate the distance it would take to stop if the car was travelling at:

i 25 mph   **ii** 35 mph   **iii** 53 mph   **iv** 68 mph.

c Use your graph to estimate how fast a car can travel and be able to stop in:

i 10 m   **ii** 25 m   **iii** 75 m.

6 A teacher asks her class to draw up a table for $y = 6 - 4x - x^2$

a This is Kylie's table.

Kylie's table

x	−3	−2	−1	0	1	2	3
6	6	6	6	6	6	6	6
−4x	+12	+8	+4	0	−4	−8	−12
−x²	+9	+4	+1	0	+1	+4	+9
y = 6 − 4x − x²	27	18	11	6	3	2	3

i Are any of Kylie's y-values correct?

ii Explain any mistakes she has made.

b This is Will's table.

Will's table

x	−3	−2	−1	0	1	2	3
6	6	6	6	6	6	6	6
−4x	−12	−8	−4	0	−4	−8	−12
−x²	−9	−4	−1	0	−1	−4	−9
y = 6 − 4x − x²	−15	−6	1	6	1	−6	−15

i Are any of Will's y-values correct?

ii Explain any mistakes he has made.

7 **a** Draw the graph of $y = 5x^2 + 4x - 8$ for $-3 \leqslant x \leqslant +2$

b Use your graph to solve the following equations.

i $5x^2 + 4x - 8 = 0$ **ii** $5x^2 + 4x - 8 = 10$ **iii** $5x^2 + 4x = 0$

c Emma says that the equation $5x^2 + 4x - 8 = -10$ cannot be solved.

i Is she correct?   **ii** Give a reason for your answer.

8 **a** Draw the graph of $y = 9 + 5x - 3x^2$ for $-2 \leqslant x \leqslant 4$

    b **i** Write down the solutions of the equation $9 + 5x - 3x^2 = 0$

      ii Explain how you found the solutions and why your method works.

    c **i** Use your graph to solve the equation $3x^2 = 5x + 6$

      ii Explain how you found the solutions and why your method works.

 9 **a** Draw the graph of $y = 3x(2 - x)$ for values of x from -1 to 3.

    b For what value of c does the equation $6x - 3x^2 = c$ have just one solution?

    c **i** Describe the values of c for which the equation $6x - 3x^2 = c$ has no solutions.

      ii Give a reason for your answer.

Homework 3

In questions **1** to **6** draw graphs for $-4 \leqslant x \leqslant +4$ to solve the simultaneous equations.

 1 $y = x + 4$
      $y = x^2$

3 $y = 3x + 1$
      $y = -x^2$

5 $y = 2x$
      $y = 3x^2 - 2$

 2 $y = 2x$
      $y = x^2 - 5$

4 $x + y = 7$
      $y = 2x^2$

6 $x - y + 4 = 0$
      $y = x(x - 1)$

 7 **a** On the same axes draw the graphs of $y = 5 + 2x - x^2$ and $y = x + 1$ for $-2 \leqslant x \leqslant +4$

    b Use your graph to solve the simultaneous equations $y = 5 + 2x - x^2$ and $y = x + 1$

    c Write down and simplify the quadratic equation whose solutions are given by the x-coordinates of the points of intersection of the graphs.

    d How can you tell from the graph that the simultaneous equations $y = 5 + 2x - x^2$ and $y = x + 7$ have no solutions?

8   **a** Draw the graph of $y = 3x^2 - 7x + 6$ for $-1 \leqslant x \leqslant +3$

     b By drawing other lines on your graph, use it to find the solutions of these pairs of simultaneous equations.

       i $y = 3x^2 - 7x + 6$ **ii** $y = 3x^2 - 7x + 6$ **iii** $y = 3x^2 - 7x + 6$

           $y = 5 - x$ $y = 3 - x$ $y = 1 - x$

 9   **a** Draw the graph of $y = x^2$ for $-4 \leqslant x \leqslant +4$

     b By drawing other lines on your graph, use it to solve these quadratic equations.

       i $x^2 + 2x = 0$ **ii** $x^2 - 2x - 1 = 0$ **iii** $x^2 + 3x - 2 = 0$

     Remember to write down the equations of your other lines.

10   **a** Complete the table of values for $y = x^2 - 3x - 3$

x	-2	-1	0	1	2	3	4
y	7		-3		-5		1

     b Draw the graph of $y = x^2 - 3x - 3$ for values of x between -2 and 4.

     c Write down the solutions of $x^2 - 3x - 3 = 0$

     d Find graphically the solutions of $x^2 - 3x - 3 = x - 6$

11   **a** Complete the table of values for $y = x^2 - 2x - 4$

x	-2	-1	0	1	2	3	4
y			-4			-1	4

     b Draw the graph of $y = x^2 - 2x - 4$ for values of x between -2 and 4.

     c Write down the solutions of $x^2 - 2x - 4 = 0$

     d By drawing an appropriate linear graph write down the solutions of
       $x^2 - 3x - 2 = 0$

 12   Trudi says that you can solve the quadratic equation $3x^2 - 7x + 1 = 0$ by finding the points of intersection of the graphs of $y = 3x^2$ and $y = 1 - 7x$ Is she correct? Give a reason for your answer.

 13   Work out which graph you would draw on the graph of $y = 4x^2$ to solve each of these equations:

     a $4x^2 - 3x - 5 = 0$ **c** $11 - 4x^2 = 3x$

     b $2x^2 + x - 3 = 0$ **d** $x^2 - 2x = 1$

14 **Get Real!**

A bus passes a layby just as a car pulls out to travel in the same direction.

The following equations give for each vehicle the distance, d metres, travelled in the next t seconds.

Bus: $d = 15t$

Car: $d = 6t + 0.25t^2$

a Draw a graph showing $d = 15t$ and $d = 6t + 0.25t^2$ for $0 \leqslant t \leqslant 40$

b Write the coordinates of the point of intersection of the graphs and explain the significance of these values in this context.

14 Quadratic functions

Homework 1

Solve the equations in questions **1** to **12**.

1 $x^2 + 10x + 9 = 0$

2 $x^2 + 12x + 35 = 0$

3 $x^2 + 9x + 18 = 0$

4 $x^2 - 5x = 0$

5 $x^2 - 7x + 10 = 0$

6 $x^2 - 25 = 0$

7 $x^2 - 3x - 10 = 0$

8 $x^2 - 6x + 9 = 0$

9 $x^2 + 2x = 15$

10 $x^2 + 6x = 7$

11 $x^2 + 6 = 5x$

12 $x(x - 4) = 12$

13 **Get Real!**

A rectangular field is 6 metres longer than it is wide.

The area of the field is 280 m^2.

Using x to represent the width of the field in metres:

a show that $x^2 + 6x = 280$

b solve this quadratic equation

c find the perimeter of the field.

> Area = 280 m^2 x

14 Ewan thinks of a number, squares it, then subtracts five times the original number.

The result is 66. What are the possible numbers that Ewan started with?

Solve the equations in questions **15** to **33**.

15 $2x^2 - 11x + 5 = 0$

16 $2x^2 - 3x - 9 = 0$

17 $3x^2 - 22x + 7 = 0$

18 $4x^2 - 18x = 0$

19 $5x^2 = 9x + 2$

20 $9a^2 - 16 = 0$

21 $39a = 8 - 5a^2$

22 $6t^2 - 31t + 33 = 0$

23 $8y^2 = 14y - 3$

24 $(2x + 3)(x - 7) = 19$

25 $\dfrac{40}{x} = x + 3$

26 $x + 3 = \dfrac{2}{x + 2}$

27 $2x + 3 = \dfrac{5}{x + 3}$

28 $2x = \dfrac{x - 1}{x + 2}$

29 $\dfrac{12}{x + 5} + \dfrac{2}{x} = 1$

30 $\dfrac{2}{x + 1} + \dfrac{3}{x + 2} = 2$

31 $\dfrac{3}{x + 1} + \dfrac{4}{2x + 1} = 4$

32 $\dfrac{9}{x - 1} - \dfrac{4}{x} = 2$

33 $\dfrac{4}{x + 3} - \dfrac{3}{x + 4} - 1 = 0$

34 Find the points at which each of the following curves meets the x-axis.
Illustrate each part with a sketch of the curve.

a $y = x^2 + 2x$ 　　　　**c** $y = 2(x^2 - 1)$
b $y = 4 + 3x - x^2$ 　　**d** $y = x^2 + 4x + 4$

35 Look at Sonya's attempt to solve a quadratic equation.
What mistakes has Sonya made?

$x^2 + 9x = 8$
$(x + 8)(x + 1) = 0$
$x = 8$ or $x = 1$

36 The quadratic functions below can all be written as $y = -x^2 + bx + c$
In each case find b and c.

a

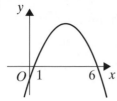

c

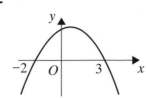

b

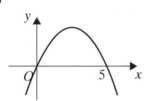

d

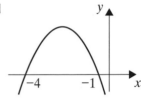

37 **Get Real!**

A rectangular paddling pool is 30 cm deep and its width is 20 cm less than its length.

The capacity of the paddling pool is 240 litres.

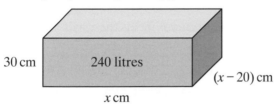

30 cm   240 litres $(x - 20)$ cm

x cm

a Show that $x^2 - 20x - 8000 = 0$

b Find the length and width of the paddling pool.

c Calculate the surface area of the paddling pool, giving your answer in square metres.

38 The sketch shows the cross-section of a wedge with its dimensions given in centimetres.

Its area is 320 cm^2.

Find the value of x.

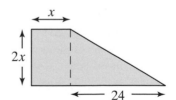

39 A rectangle is 5 cm longer than it is wide and its diagonals are 25 cm long.

Find the perimeter and the area of the rectangle.

Homework 2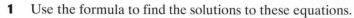

1 Use the formula to find the solutions to these equations.

    a $x^2 + 10x + 9 = 0$ **b** $x^2 - 5x = 0$ **c** $x^2 - 7x + 10 = 0$

Solve the equations in questions **2** to **16** using the formula.
Give your answers to 2 decimal places.

2 $x^2 + 8x + 3 = 0$ **7** $2x^2 = 11 - 3x$ **12** $16h - 5 = 7h^2$

3 $x^2 - 7x + 5 = 0$ **8** $6y^2 - 5y = 2$ **13** $a(5a - 3) = 22$

4 $x^2 + 9x - 1 = 0$ **9** $4p = 9 - 2p^2$ **14** $t(t - 7) - 2 = 3t^2$

5 $3x^2 - 4x - 5 = 0$ **10** $1.2x^2 - 2.5x + 0.8 = 0$ **<u>15</u>** $(y - 5)^2 = y(2y + 4)$

6 $4x^2 - x = 7$ **11** $3.5y^2 + 6.1y - 5.7 = 0$ **<u>16</u>** $\dfrac{7}{3x - 1} - \dfrac{1}{x + 4} = 1$

17 Find the points at which each of these curves meets the x-axis.

    a $y = 2x^2 + x - 7$ **b** $y = 10 - x - x^2$ **c** $y = 5x^2 + 4x - 8$

    Check your answers using the graphs you plotted for questions **4**, **3** and **7** in Chapter **13** Homework **2**.

18 Here is Kevin's attempt to use the quadratic formula to solve the equation $3 - 7x - x^2 = 0$

    Is Kevin correct?

    Give a reason for your answer.

$$x = \frac{-7 \pm \sqrt{(49 - 4 \times 3 \times -1)}}{2 \times 3} = \frac{-7 \pm \sqrt{37}}{6}$$

$$x = \frac{-7 \pm 6.0827\ldots}{6} = -8.01 \text{ or } -7.10$$

19 **Get Real!**

    A rectangular sports ground is 15 metres longer than it is wide.

    Its area is 2800 m^2.

    Area = 2800 m^2

    x

    Using x metres to represent the length of the sports ground:

    a show that $x^2 - 15x = 2800$

    b find the dimensions of the sports ground to the nearest metre.

20 Get Real!

The area of a rectangular gate is 1.74 m² and its width is 0.25 m less than its height.

Find the dimensions of the gate.

21 Get Real!

Four congruent rectangular tiles fit exactly around a square tile as shown in the diagram.

The total area is 90 cm².

a Show that $4x^2 + 20x - 65 = 0$

b Hence find the value of x.

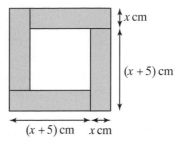

x cm

$(x + 5)$ cm

$(x + 5)$ cm x cm

22 The length of the diagonals of a square are 7 cm longer than its sides.

Find the perimeter and area of the square to 3 significant figures.

23 Get Real!

A drinks can is in the shape of a cylinder of height 14 cm.

a Explain why the total surface area of the can is given by $A = 2\pi r^2 + 28\pi r$

b The surface area of the can is 320 cm².
Find its diameter.

r

14 cm

24 Find the solutions of each of these quadratic equations.
Give exact answers using surds.

a $x^2 + 6x + 7 = 0$ **c** $x^2 - 2x - 6 = 0$ **e** $3x^2 - 6x + 1 = 0$

b $x^2 - 10x + 11 = 0$ **d** $2x^2 + 7x + 2 = 0$ **f** $5x^2 - 4x - 2 = 0$

25 Get Real!

A boy runs a distance of 10 kilometres, then walks for 5 kilometres.
His average walking speed is 2 km/h slower than his average running speed and the total journey takes 2 hours 15 minutes.

a Show that $\dfrac{10}{x} + \dfrac{5}{x - 2} = \dfrac{9}{4}$ where x km/h is his average running speed.

b Solve this equation to find his average running speed.

Homework 3

In questions **1** to **3**, write the equation in the form $(x + a)^2 + b = 0$ and hence find the solutions of the equation to 2 decimal places.

1 $x^2 + 4x - 2 = 0$ **2** $x^2 + 10x - 12 = 0$ **3** $x^2 + 6x + 2 = 0$

In questions **4** to **6**, write the equation in the form $(x - a)^2 + b = 0$ and hence find the solutions of the equation to 2 decimal places.

4 $x^2 - 2x - 7 = 0$ **5** $x^2 - 8x + 3 = 0$ **6** $x^2 - x - 5 = 0$

In questions **7** to **9**, write the equation in the form $(x + a)^2 = b$ and hence find the solutions, giving your answers in surd form.

7 $x^2 + 2x = 5$ **8** $x^2 + 8x = 7$ **9** $x^2 + x = 3$

In questions **10** to **12**, write the equation in the form $(x - a)^2 = b$ and hence find the solutions, giving your answers in surd form.

10 $x^2 - 6x = 11$ **11** $x^2 - 4x = 1$ **12** $x^2 - 5x = -2$

In questions **13** to **18**:

   i   write the function in the form $y = (x + a)^2 + b$ where a and b are positive or negative constants

   ii   find the coordinates of the lowest point on the curve

   iii   sketch the curve and show its line of symmetry.

13 $y = x^2 + 4x + 7$ **15** $y = x^2 - 10x + 27$ **17** $y = x^2 + 7x$

14 $y = x^2 - 2x - 4$ **16** $y = x^2 + 6x + 1$ **18** $y = x^2 - 5x + 8$

19   Rita has been asked to write $x^2 - 9x + 3$ in the form $(x - a)^2 + b$

Here is her working. Is Rita correct?

Give reasons for your answer.

$(x - a)^2 + b = x^2 - a^2 + b$
If this $= x^2 - 9x + 3$,
then $b = 3$
and $a^2 = 9$ so $a = 3$

20 **a** Find a and b such that $3x^2 - 12x + 7 = 3(x - a)^2 + b$ where a and b are positive or negative.

    b Use your answer to part **a** to:

      i solve the equation $3x^2 - 12x + 7 = 0$

      ii find the lowest point on the graph of $y = 3x^2 - 12x + 7$ and sketch it.

 21 **a** Find a and b such that $5 - 10x - x^2 = b - (x + a)^2$.

    b Use your answer to part **a** to:

      i solve the equation $5 - 10x - x^2 = 0$

      ii find the highest point on the graph of $y = 5 - 10x - x^2$ and sketch it.

 22 A right-angled triangle has sides of length x cm, $(x + 2)$ cm and $(x - 4)$ cm as shown.

    a Show that $x^2 - 12x + 12 = 0$

    b Find the solutions of this equation by completing the square.

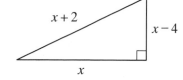

    c Write down the lengths of the sides of the triangle to the nearest millimetre.

 23 **Get Real!**

The sketch shows the cross-section of a summer house with its dimensions in metres.

    a If the area of the cross-section is 6.56 m^2, show that $(x + 2.2)^2 = 11.4$

    b Hence find the value of x to 2 decimal places.

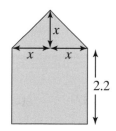

15 Inequalities and simultaneous equations

Homework 1

1 Write the inequalities shown on these diagrams.

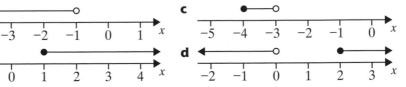

2 List all the integer values of t such that:

a $-10 < 5t \leqslant 7$ **b** $-9 \leqslant 3t < 3$ **c** $-3 < 2t + 3 \leqslant 5$

3 Find the smallest integer n such that $2n + 7 \leqslant 4n - 2$

4 Solve these linear inequalities, showing your solutions on a number line.

a $2x - 3 \geqslant x + 9$ **c** $\dfrac{x}{4} + \dfrac{2x - 5}{6} < 5$

b $2(3 - 4x) > 1 - 6x$ **d** $\dfrac{x - 2}{3} \leqslant x - \dfrac{2(x + 1)}{5}$

5 Mrs Reynolds says to her class, 'I am thinking of an integer. When I double it and add 5, the result is more than the original number plus 1.'

a Write the problem as an inequality.

b What is the smallest number that Mrs Reynolds is thinking of?

6 The integers x and y are such that $-3 < x \leqslant 2$ and $-2 \leqslant y < 1$

a What is the smallest value of x^2?

b What is the largest value of y^2?

c What is the largest value of xy?

Homework 2

1   On separate diagrams draw the regions defined by the following inequalities.

   a $x \leqslant 3$ **b** $y > -2$ **c** $x < -1$ **d** $0 \leqslant y < 4$

2   **a** On the same graph draw and label the regions:

   i $-2 \leqslant x < 4$ **ii** $-3 < y \leqslant 2$

   b Clearly label the region for which both sets of inequalities are true.

   c Are any of these points in this region?

   i $(-2, -3)$ **ii** $(2, -3)$ **iii** $(3, 2)$ **iv** $(-2, 2)$

3   **a** Draw axes for x and y from -5 to 5 with equal scales on both axes.

   b Draw the lines with equations $2y = x + 4$, $2y + x + 2 = 0$ and $x = 4$

   c Shade the region on the graph where $2y \leqslant x + 4$, $2y + x + 2 \geqslant 0$, $x \leqslant 4$ and $x \geqslant 0$

   d What is the shape of this region?

<u>4</u>

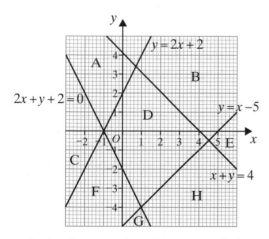

   a In the diagram state which region is bounded by the lines described by:

   i $y \geqslant x - 5$, $2x + y + 2 \leqslant 0$, $y \leqslant 2x + 2$

   ii $y \leqslant x - 5$, $2x + y + 2 \geqslant 0$, $x + y \leqslant 4$

   iii $y \geqslant 2x + 2$, $2x + y + 2 \geqslant 0$, $x + y \leqslant 4$

   b Describe with inequalities the regions:

   i B   **ii** C   **iii** E and H   **iv** C and F   **v** B and D and F

5 **a** Draw a grid for x and y from 0 to 15.

b On your grid indicate clearly the region defined by the four inequalities:

$x \geqslant 1$

$y \geqslant 5$

$3x + y \geqslant 12$

$x + y \leqslant 15$

c If you also know that x and y are integers and $y = 9x$, what point must (x, y) be?

Homework 3

 1 Solve these pairs of simultaneous equations.

a $7x + 2y = 26$ **c** $3x + 2y = 18$ **e** $7x + 4y = 9$

     $3x - 2y = 14$ $5x + 4y = 32$ $6y = 19 - 5x$

b $4x + y = 17$ **d** $7x - 2y = 8$ **f** $5x - 4y = 14$

      $x = 3 + y$ $3x = 9 - y$ $6x - 7y = 19$

 2 **Get Real!**

Julie buys seeds for her garden by mail order.

She pays £9.66 for 3 packets of sunflower seeds and 1 packet of marigold seeds.

A packet of sunflower seeds costs 70 pence more than a packet of marigold seeds.

Write two simultaneous equations in x and y, where x is the cost of a packet of sunflower seeds and y is the cost of a packet of marigold seeds.

Solve your equations to work out how much packets of sunflower seeds and marigold seeds cost.

 3 **Get Real!**

Jess buys 3 portions of fish and 2 portions of chips from the Fish Shop and she pays £8.25

Barney buys 2 portions of fish and 5 portions of chips and he pays £9.35

How much does a portion of fish and chips cost?

4 Get Real!

A travel brochure quotes the price of a holiday to Spain as £900 for 2 adults and 3 children.

The Monaghan family consists of 3 adults and 4 children.

They are quoted £1280 for the same holiday.

What is the cost of the holiday for each adult and each child?

 5 Get Real!

A company that fits kitchens and bathrooms currently has 12 kitchen displays and 5 bathroom displays in its showroom.

The displays cover an area of 98 square metres.

The show room is being redesigned and will have 10 kitchen displays and 7 bathroom displays.

The displays will then cover an area of 93 square metres.

Find the area of a kitchen display and of a bathroom display.

6 Get Real!

Last month Anil sent 77 texts and made 180 minutes of calls.

His mobile phone bill was £11.31

This month he has sent 102 texts and made 270 minutes of calls.

His bill this month is £16.56

How much is a text and how much does he pay per minute for his calls?

Homework 4

 1 Sophie is trying to solve the simultaneous equations $y = 2x^2 - x$ and $y = 12x + 7$

She draws this table.

x	0	1	2	3	4	5	6	7	8
$2x^2 - x$	0	1	6						
$12x + 7$	7	19	31						

a Copy and complete Sophie's table.

Have you found a solution of the simultaneous equations?

Now solve the simultaneous equations using the substitution method.

b Was the table a good method to use? Give a reason for your answer.

2 Solve these pairs of simultaneous equations.

a $y = 3x^2$ **c** $y = x^2 + 2x - 1$ **e** $x^2 + y^2 = 10$

   $y = 2 - 5x$ $y = 6x + 20$ $2x + y = 5$

b $y = 9x - 4$ **d** $y = 4x^2 + 2x$ **f** $x^2 + y^2 = 289$

   $y = 5x^2$ $6 = 3x + y$ $y = 2x - 1$

3 Dylan is designing a children's slide.

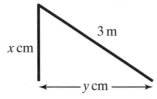

He wants the vertical height of the slide to be 60 centimetres more than half the horizontal base length.

a If the base length is y cm and the height is x cm, form an equation in x and y.

The straight part of the sloping slide is 3 metres long.

b Use Pythagoras' theorem to state a second equation in x and y.

c Solve your equations simultaneously to find the base length and height of Dylan's slide.

16 Trigonometry

Homework 1

1 Find the length of the sides marked x in these triangles.
Give your answers to an appropriate degree of accuracy.

a
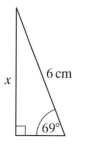

b
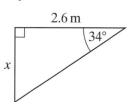

c
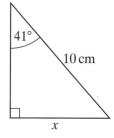

d
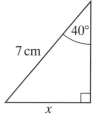

e
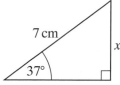

f

2 Calculate the lengths marked x in these triangles.
Give your answers to an appropriate degree of accuracy.

a
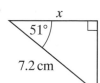

b
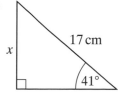

c
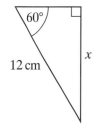

3 A *clinometer* is an instrument which is used to measure the height of tall buildings. It measures the angle of elevation.

The diagram below illustrates a clinometer in use.

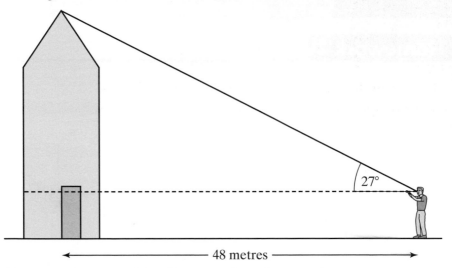

48 metres

The surveyor's eye level is 1.73 metres above the ground.

a Draw an appropriate right-angled triangle, and label the opposite side, the adjacent side and the hypotenuse.

b Calculate the height of the building above the surveyor's eye level.

c What is the height of the building?

4 Find the length of the hypotenuse in each of these triangles.

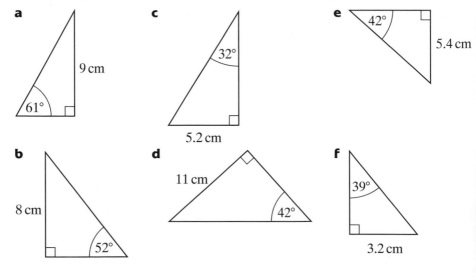

5 Calculate the sides marked x in these triangles.

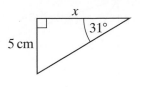

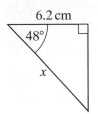

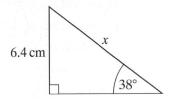

Homework 2

1 Calculate the size of the angles marked x in these triangles.

a
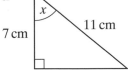

b

c
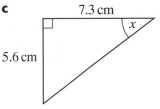

2 Calculate the size of the angles marked x in these triangles.

a
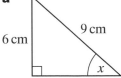

c
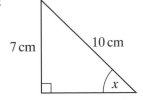

e
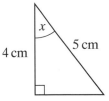

b
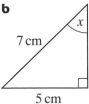

d

f
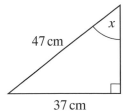

3 The line AB is 7 cm long, and at right angles to the line BC which is 13 cm long.

Sketch the triangle ABC, and calculate the size of all the angles in the triangle and the length of the side AC.

<u>4</u> In the diagram, AB = 8 cm, AD = 9 cm, and angles ABC and ADC are both right angles.

  a Calculate x, the length of AC.

  b Calculate y, the size of angle ACD.

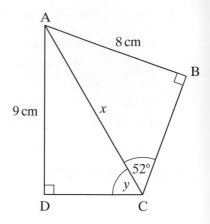

Homework 3

1 In this cuboid, GH = 8 cm, CG = 6 cm and FG = 5 cm.

  a Calculate angle DGH.

  b Calculate angle GBH.

  c Calculate angle FDH.

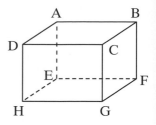

2 The diagram shows an isosceles triangular prism.

ED = DF = 9 cm,

FC = 7 cm and angle EFD = 71°.

  a Calculate the height of the prism.

  b Calculate the size of angle AFC.

  c Calculate the angle AF makes with the horizontal.

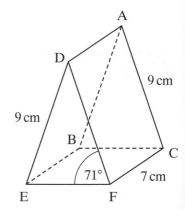

3 The diagram shows a cylinder with diameter 8 cm and height 7 cm underneath a cone with a base of diameter 8 cm and a height of 6 cm.

BC and DE are diameters with BC vertically above DE.

a Calculate the size of the angle ABC.

b Calculate the size of the angle ADE.

c Calculate the length AD.

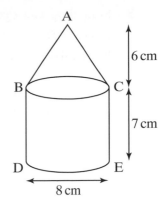

4 The diagram shows a barn roof.

CDEF is a rectangle measuring 9 m × 12 m.

ACF and BDE are congruent isosceles triangles with a base of 9 m.

ABEF is an isosceles trapezium.

AB = 8 m and the angle BEF = 72°.

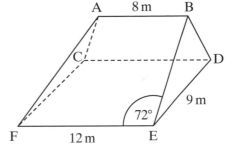

a Calculate the length of BE.

b Calculate the height of the isosceles trapezium ABEF.

c Calculate the angle BED.

d Calculate the height of the roof.

Homework 4

1 Sketch the graph of $y = \sin x$ for values of x from 0° to 360°.

Use your sketch to find out which angles between 0° and 360° have the same sine as:

a 80°   **c** 99°   **e** 342°   **g** 370°

b 70°   **d** 222°   **f** 212°   **h** −10°

2 Sketch the graph of $y = \cos x$ for values of x from 0° to 360°.

Use your sketch to find out which angles between 0° and 360° have the same cosine as:

a 20°   **c** 33°   **e** 234°   **g** 345°

b 70°   **d** 101°   **f** 500°   **h** −50°

3 Solve these equations, giving all the answers in the range $0°$ to $360°$.

a $\sin x = 0.8$ **c** $\sin x = 0.55$ **e** $\cos x = 0.12$

b $\cos x = 0.2$ **d** $\cos x = -0.43$ **f** $\sin x = -0.25$

4 Solve these equations, giving all the answers in the range $0°$ to $360°$.

a $5\sin x = 2$ **c** $12\cos x = -4$ **e** $12\sin x = -7$

b $3\cos x = 1$ **d** $7\cos x = -2$ **f** $6\sin x + 1 = 0$

 5 **a** Sketch a graph of $y = \cos x$ for $0° \leqslant x \leqslant 360°$.

b Use your sketch and a calculator to solve these equations to one decimal place.

i $\cos x = 0.33$ **iii** $\cos x = 0.25$ **v** $\cos x = -\dfrac{\sqrt{3}}{2}$ **vii** $\cos x = -\dfrac{1}{\sqrt{2}}$

ii $\cos x = 0.87$ **iv** $\cos x = \dfrac{\sqrt{3}}{2}$ **vi** $\cos x = \dfrac{1}{\sqrt{2}}$ **viii** $\cos x = -1.5$

 6 Sketch a graph of $y = \cos x + 5$ for $0° \leqslant x \leqslant 360°$.

a Explain how to obtain the graph of $y = \cos x + 5$ from the graph of $y = \cos x$.

b What is the maximum value of y? For what values of x does it occur?

c What is the minimum value of y? For what value of x does it occur?

d What is the range of values of y?

e For what values of k can the equation $\cos x + 5 = k$ be solved?

f Explain the difference between the graph of $y = \cos x + 5$ and the graph of $y = \cos (x + 5)$.

 7 **a** Sketch a graph of $y = -\sin x$ for $0° \leqslant x \leqslant 360°$.

b What are the similarities and differences between this graph and the graph of $y = \sin x$?

c Which of these statements are true?

i $y = -\sin x$ crosses the x-axis at the same points as $y = \sin x$.

ii $y = -\sin x$ crosses the y-axis at the same point as $y = \sin x$.

iii The range of values of y is the same as for $y = \sin x$ but negative.

iv All the y-values have the opposite sign from those for $y = \sin x$.

8 Repeat question **7** for the graph of $y = -\cos x$.

9 **Get Real!**

The diagram represents the London Eye, the world's biggest Ferris wheel.

The radius of the wheel is approximately 65 metres.

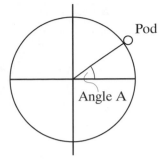

As the wheel makes a complete revolution, the angle A increases from 0° at the start when the pod is level with the centre of the wheel to 360° when the pod returns to its original position.

a For different values of the angle A, work out the height of the pod above the centre of the wheel.

b Copy and complete this table to show the height of the pod for different values of angle A.

Angle (degrees)	0	30	60	90	120	150	180	210	240	270	300	330	360
Height (m)							0						

c Draw a graph of these values, with angle on the horizontal axis and height on the vertical axis.

d Comment on the shape of your graph.

e Find the equation of the graph.

Homework 5

1 Calculate the length of the sides marked x in these triangles.

a
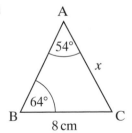

b
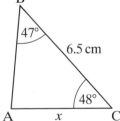

c
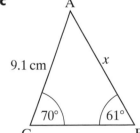

2 Calculate the size of the angles marked x in these triangles.

a

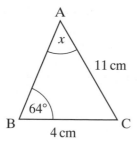

c

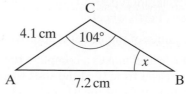

b

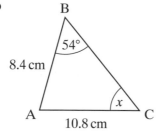

3 For each part, draw and label a diagram, and use the sine rule to find the sides and angles marked '?'.

	A	B	C	a	b	c
a	53°	80°		10.4 cm	?	
b	?		71°	8 cm		10.3 cm
c	?	114°		8 cm	17 cm	
d	71°		88°	12.3 cm	?	
e	?		75°		7.1 cm	7.4 cm

4 The bearing of Brome from Croft is 130°.

Dwight is 4.6 km from Brome on a bearing of 071°.

The distance from Croft to Dwight is 7.1 km.

Calculate the bearing of Dwight from Croft.

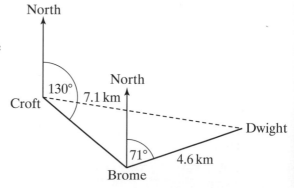

5 A triangle ABC has AB = 9 cm, BC = 7.4 cm and angle A = 47°.
Calculate the two possible values of angle C.

Homework 6

1 Calculate the lengths of the sides marked *x* in these triangles.

a
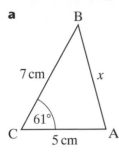

b
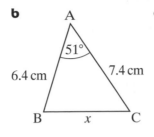

c
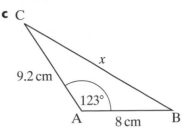

2 Calculate the size of the angles marked *x* in the these triangles.

a
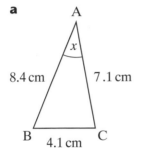

b
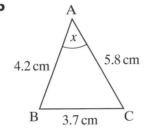

c
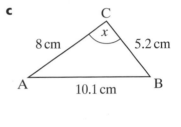

3 For each part, draw and label a diagram, and use the cosine rule to find the sides and angles marked '?'.

	A	B	C	a	b	c
a		74°		7 cm	?	8 cm
b	?			5.6 cm	7.1 cm	10.3 cm
c	?			13 cm	7.6 cm	8.2 cm
d			77°	7.7 cm	8.5 cm	?
e			150°	4.2 cm	3.8 cm	?

4 The village of Istle is 12 km from Alton on a bearing of 114°.

Onber is 8.4 km from Istle, on a bearing of 071°.

How far is it from Alton to Onber?

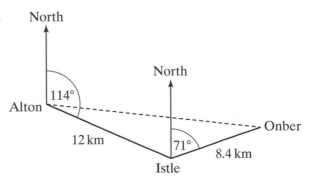

<u>5</u> **Proving the cosine rule**

Copy and complete these steps to prove the cosine rule.

1 Label the sides a, b and c.

2 If BD $= x$, write CD in terms of a and x.

> CD =

3 Use Pythagoras' theorem on the left-hand triangle.

> $c^2 =$

4 Rearrange **3** to make h^2 the subject.

> $h^2 =$

5 Use Pythagoras' theorem on the right-hand triangle.

> $b^2 = ?^2 + (?)^2$

6 Remove the brackets ... carefully. Watch out for the signs.

> $b^2 =$

7 Rearrange this to make h^2 the subject.

> $h^2 =$

8 Use your answers to **4** and **7** to write an equation without h.

> $c^2 - x^2 =$

9 Rearrange **8** to make b^2 the subject.

> $b^2 =$

10 Write $\cos B$ in terms of c and x.

> $\cos B =$

11 Rearrange **10** to make x the subject.

> $x =$

12 Rewrite **9**, replacing x with your answer to **11**.

> $b^2 =$

This is the cosine rule.

Homework 7

1 Calculate the areas of these triangles.

a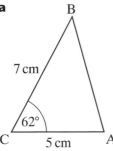

c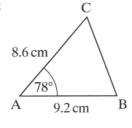

e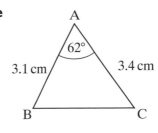

b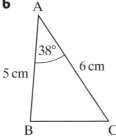

d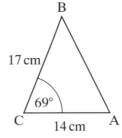

f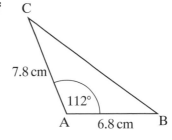

2 Calculate the area of triangle ABC.

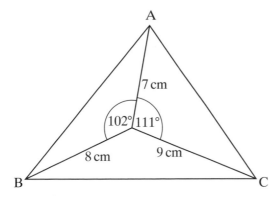

3 A triangle has an area of 26 cm².
One side is 8 cm long and another side is 9 cm long.
Calculate the possible sizes of the angle between these two sides.

4 These three triangles all have the same area. Calculate angle x and length y.

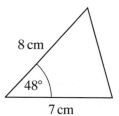
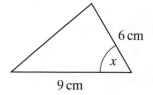
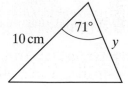

Homework 1

Apart from question 5 this is a non-calculator exercise.

1   For this question, use the graph of $y = x^3 - x^2 - 2x$

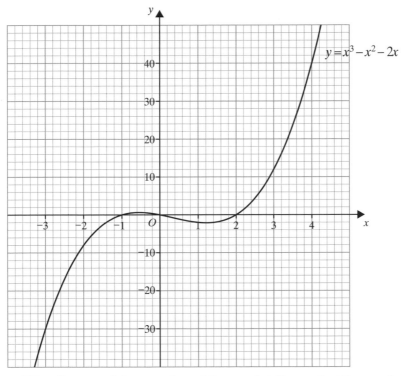

$y = x^3 - x^2 - 2x$

a What feature of the equation of the graph tells you that the graph goes through the point $(0, 0)$?

b Use the graph to solve the equations:

   i $x^3 - x^2 - 2x = 0$ **ii** $x^3 - x^2 - 2x = 10$ **iii** $x^3 - x^2 - 2x + 5 = 0$

c Find the range of values of k for which the equation $x^3 - x^2 - 2x = k$ has more than one solution.

d Solve the equations:

   i $x^3 - x^2 - x = 0$ **ii** $x^3 - x^2 - 3x = 0$

2 On one diagram, draw the graphs of $y = x^3$ and $y = x^3 - 5$ for values of x from -3 to 3.

   a What is the same and what is different about the two graphs?

   b Draw the line $y = -3$ on the diagram and use it to solve the equations:

     i $x^3 = -3$ **ii** $x^3 - 5 = -3$

   c Solve the equations in **b** algebraically.

3 For this question, use the graph of $y = -x^3 + 5x^2 + 10x - 10$ below.

   a Compare this graph with the graph in question **1**.

     What feature of the equation $y = -x^3 + 5x^2 + 10x - 10$ indicates that the graph:

     i does not pass through $(0, 0)$

     ii is 'upside down' compared with the graph in question **1**?

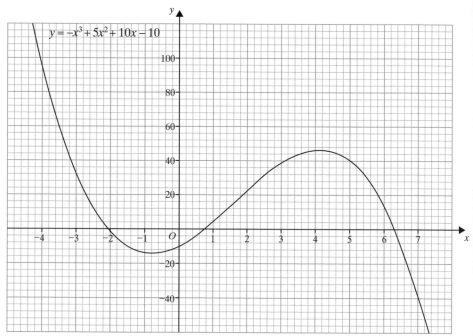

   b Solve the equations:

     i $-x^3 + 5x^2 + 10x - 10 = 0$

     ii $-x^3 + 5x^2 + 10x - 30 = 0$

     iii $-x^3 + 5x^2 + 10x - 10 = x$

   c Find another equation that has only one solution.

   d Find another equation that has three solutions.

   e Find an equation that has two solutions.

4 **a** Copy and complete this table of values for the function $y = x^3 - 4x + 2$
for $-4 \leqslant x \leqslant 4$

x	-4	-3	-2	-1	0	1	2	3	4
x^3	-64								
-4x	16								
2	2	2	2	2	2	2	2	2	2
y	-46								

  b Draw suitable axes on graph paper, plot the points and join them with a smooth curve.

  c i Write the x-coordinates of the points where the graph cuts the x-axis.

   ii Write the equation you have just solved.

  d Use the graph to solve the equation $x^3 - 4x + 2 = 3$

  e Solve the equation $x^3 - 4x = 0$

   i by drawing a line on the graph

   ii by using algebra.

5 **Get Real!**
A rectangular box is constructed from a rectangle of card measuring
30 cm by 50 cm. A square measuring x cm by x cm is cut from each
corner and the resulting flaps are folded up to make the sides of the box.

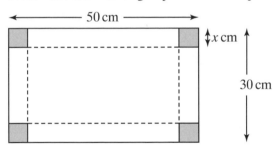

  a Explain why the volume of the box, V cm^3, is given by the cubic function
$V = x(30 - 2x)(50 - 2x)$

  b i Use the equation $V = x(30 - 2x)(50 - 2x)$ to show that the values of x for
which $V = 0$ are 0, 15 and 25.

   ii Use the diagram to explain why the volume is zero when the length of
the sides of the cut out squares is 0 cm or 15 cm or 25 cm.

  c i Draw the graph of the function $y = x(30 - 2x)(50 - 2x)$ for values of x
from 0 to 15 (you may like to use a graphic calculator or a computer
graph-drawing program).

   ii Use the graph to find the maximum possible volume of the box and the
value of x that gives the maximum volume.

6   Get Real!

In Apply **1**, the cubic equation $y = -\dfrac{x^3}{3000} + \dfrac{3x^2}{40} - 5x + 120$, where the horizontal distance is x feet and the corresponding height is y feet, was used as a model of part of a roller coaster ride.

a i   Use the equation $y = -\dfrac{x^3}{3000} + \dfrac{3x^2}{40} - 5x + 120$ to find the difference in height between the bottom of the dip and the top of the rise.

ii   Show how to use the graph to check that your answer is reasonable.

iii   Use the graph to find the horizontal distance between the bottom of the dip and the top of the rise.

b   What is the effect on the graph of:

i   changing the number 120

ii   changing $-\dfrac{x^3}{3000}$ to $\dfrac{x^3}{3000}$?

7 Beyond cubic functions, with the general form $f(x) = ax^3 + bx^2 + cx + d$ are the quartic functions and the quintic functions. The simplest versions of these are $y = x^4$ and $y = x^5$. The diagram shows the graphs of $y = x$, $y = x^2$, $y = x^3$, $y = x^4$ and $y = x^5$

Identify which graph is which and explain how you made your decisions.

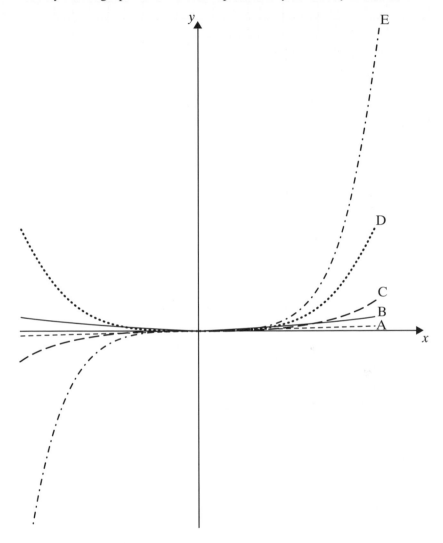

Homework 2

 1 **a** Copy and complete this table of values for the reciprocal function $y = \dfrac{6}{x}$

x	−6	−5	−4	−3	−2	−1	0	1	2	3	4	5	6
y			−1.5				−		3				

b Draw the graph of the function.

c Use the graph to solve the equations:

   i $\dfrac{6}{x} = 5$ **ii** $\dfrac{6}{x} = -5$ **iii** $\dfrac{6}{x} = x$ **iv** $\dfrac{6}{x} = -x$

 2 For this question, use the graph of the reciprocal function $y = \dfrac{5}{x} - 3$ below.

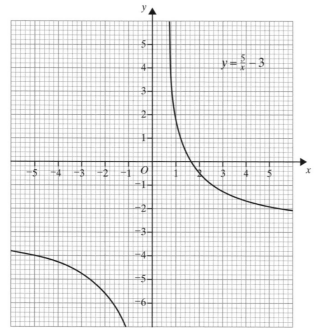

$$y = \frac{5}{x} - 3$$

a Use the graph to solve the equations:

   i $\dfrac{5}{x} - 3 = 2$ **ii** $\dfrac{5}{x} - 3 = x$ **iii** $\dfrac{5}{x} - 3 = 4x$

b **i** Find an equation that has no solutions.

   ii Is it possible to find equations with: one solution; two solutions; more than two solutions? Use diagrams to explain your answers.

c Solve the equations in part **a**, using algebra instead of the graph. (You will need a calculator.)

   Give your answer to an appropriate degree of accuracy.

3 Here is the table of values for the equation $y = \dfrac{6}{x} + 6$

x	-10	-8	-6	-4	-2	0	2	4	6	8	10
$\dfrac{6}{x}$	-0.6					$-$	3				0.6
$+6$	6	6	6	6	6	6	6	6	6	6	6
y						$-$	9				

a Copy the table and complete it.

b Plot the points with the x-coordinates and y-coordinates in the table and join the two separate parts of the graph with smooth curves.

c Where does the graph cut the x-axis? Show how to find the answer:

   i using the graph   **ii** using algebra.

d How does the graph of the equation $y = \dfrac{6}{x}$ compare with the graph of the equation $y = \dfrac{6}{x} + 6$?

e Use the graph of $y = \dfrac{6}{x} + 6$ to solve the equations:

   i $\dfrac{6}{x} + 6 = 10$ **ii** $\dfrac{6}{x} + 6 = x$ **iii** $\dfrac{6}{x} - x + 1 = 0$

f Solve the equations in **e** algebraically.

   Give your answer to an appropriate degree of accuracy.

4 **a** On one diagram, draw the three graphs with equations $y = \dfrac{1}{x}$, $y = \dfrac{4}{x}$ and $y = \dfrac{8}{x}$

b Write what is the same and what is different about the graphs.

c Without drawing it, explain what the graph of $y = \dfrac{12}{x}$ would be like in relation to the other three graphs.

5 On the same diagram, draw the graphs of $y = \dfrac{10}{x^2}$ and $y = -\dfrac{10}{x^2}$ for values of x from -10 to 10.

a What are the similarities and what are the differences between the two graphs?

b What would the graph of $y = \dfrac{10}{(-x)^2}$ be like?

6 A rectangle has an area of 50 cm². Its width is x cm and its height is y cm.

a Which of these expressions correctly express the relationship between x and y?

i $\dfrac{x}{y} = 50$ **iii** $\dfrac{50}{x} = y$ **v** $\dfrac{y}{50} = x$

ii $\dfrac{y}{x} = 50$ **iv** $\dfrac{50}{y} = x$ **vi** $\dfrac{x}{50} = y$

b Draw a graph of y against x for $0 < x \leqslant 50$

c Use the graph to find:

  i the measurements of the rectangle when it is a square

  ii the measurements of the rectangle when its length is twice its width.

d Solve these equations algebraically, showing all the steps of your method, and check that your answers are the same as those obtained from the graph.

 7 A rectangular box with a square base has a volume of 1000 cm³. The edges of the base are x cm long and the height of the box is y cm.

a Find an equation expressing y in terms of x.

b Use your equation to find the height of the box when its base measures 20 cm by 20 cm.

c i Draw a graph of y against x for appropriate values of x and y.

  ii Use your graph to find the dimensions of the box when x and y are equal.

d Which of these statements are true?

i $x = \sqrt{\dfrac{1000}{y}}$ **ii** $y \propto \dfrac{1}{x}$ **iii** $x \propto \dfrac{1}{y^2}$ **iv** $x^2 y = 1$

 8 **Get Real!**

An expedition has enough food for 12 people for 4 days.

a If two people drop out of the trip, how long will the food last?

b If the expedition only lasts 2 days, how many people can go?

c If x is the number of people on the expedition and y is the number of days:

  i show that y is a reciprocal function of x

  ii sketch a graph of y against x.

Homework 3

 1 Work out:

   a $10^{-2}, 10^{-1}, 10^0, 10^1, 10^2$ **d** $0.1^{-2}, 0.1^{-1}, 0.1^0, 0.1^1, 0.1^2$

   b $5^{-2}, 5^{-1}, 5^0, 5^1, 5^2$ **e** $0.2^{-2}, 0.2^{-1}, 0.2^0, 0.2^1, 0.2^2$

   c $3^{-2}, 3^{-1}, 3^0, 3^1, 3^2$

2 The equation $y = 1.5^x$ represents an exponential function.

   a Use the equation to find:

      i the value of y when x is 3

      ii the value of x for which y is 10 (you will need to use the trial and improvement method for this)

      iii the point where the graph of $y = 1.5^x$ cuts the y-axis.

   b Sketch a graph of $y = 1.5^x$

3 The graph shows a model of the growth of a culture of bacteria, the exponential function $y = 10 \times 1.02^x$, where y grams is the mass of bacteria x minutes the start of the experiment.

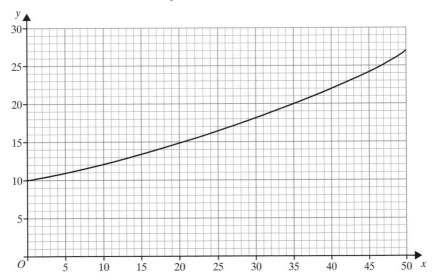

   a Use the graph to estimate:

      i the mass of bacteria after 5 minutes

      ii the mass of bacteria at the start of the experiment

      iii the time taken for the mass of bacteria to double.

   b What percentage increase in the mass of the bacteria is there:

      i after 1 minute

      ii after 10 minutes?

111

4 Under certain economic situations, house prices can rise exponentially. For example, house prices could rise by 20% a year for a period of time.

a Find a formula for the price of a house initially costing £50 000 whose price is increasing by 20% a year. Define any letters that you use to represent the quantities in your formula.

b Use your formula to calculate the price of the house after 5 years.

c Why might someone expect the price after 5 years to be £100 000?

d Use the trial and improvement method to find how long it would take for the price of the house to increase from £50 000 to £200 000.

e Explain why house prices may not fit this pattern.

 5 The general form of exponential functions is $f(x) = Ak^{mx}$

Identify the values of A, k and m in each of these exponential functions:

a $f(x) = 100 \times 1.1^x$ **b** $f(x) = 3 \times 10^{-x}$ **c** $f(x) = \dfrac{1}{2^{3x}}$

 6 This is a sketch of the graph of $y = 2^x$ for values of x from -1 to 5.

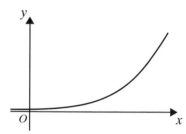

a What is the value of y when x is 0?

b Copy the diagram and sketch on it the graph of these equations:

i $y = 3^x$ **ii** $y = \dfrac{1}{2}^x$ **iii** $y = 1.5^x$ **iv** $y = -2^x$

 7 £4500 is invested in a bank account paying 3.2% a year interest; the interest is added on at the end of each year.

a If £y is the amount in the account after x years, find a formula connecting x and y.

b Use the formula to calculate:

i the amount in the account after 3 years

ii the amount in the account after 5 years.

c Draw a graph of y against x for $0 \leqslant x \leqslant 10$

d Use the graph to estimate how long it will take for the amount in the account to exceed £5000.

8 A radioactive isotope of carbon, Carbon 14, is used in carbon dating. The carbon loses mass very slowly by radioactive decay. If the mass of Carbon 14 initially is 1 kg, the approximate mass, y kg, remaining after x thousand years is given by the equation $y = 0.886^x$

a Use this formula to find the mass of Carbon 14 remaining after 10 thousand years.

b Use the trial and improvement method to find how many years it takes for the mass remaining to be less than 0.5 kg.

c Draw the graph of $y = 0.886^x$ for $0 \leqslant x \leqslant 20$

d Use your graph to show that the time taken for the mass of Carbon 14 to halve from 1 kg to 0.5 kg is the same as the time taken to halve from 0.5 kg to 0.25 kg.

e By what percentage does the mass of Carbon 14 decrease over 5000 years?

18 Loci

Homework 1

1 Get Real!

  a Draw a sketch of the locus of the girl as she goes up and down on the seesaw.

  b How would the locus of the boy be different?

2 AB is perpendicular to BC. Draw the locus of points that are twice as far from BC as they are from AB.

> **HINT**
> Imagine AB is the y-axis of a coordinate grid, and BC is the x-axis. Can you find the coordinates of some points twice as far from the x-axis as from the y-axis?

3 Mark two points, A and B, 8 cm apart.

Find two points that are 3 cm from A and 7 cm from B.

Find two points that are 4 cm from A and 6 cm from B.

Find some more points where the total distance from A and B is 10 cm.

Join them up to draw the locus of all points with a total distance of 10 cm from A and B.

<u>4</u> Get Real!

Match the loci below with the descriptions that follow.

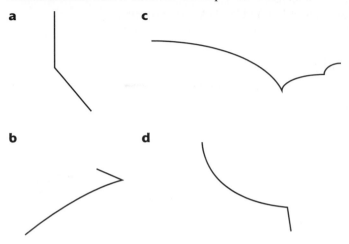

i A football hitting a wall

ii A boy going down a slide and falling off the bottom

iii A man jumping out of an aeroplane and opening a parachute

iv A cricket ball being bowled, bouncing once and then being hit

5 Get Real!

A garden is 12 m long and 8 m wide.

a Using a scale of 1 cm to 2 m, make a scale drawing of the garden.

b The owner builds a patio, 2 m wide, along the edge AD. Draw this on your plan.

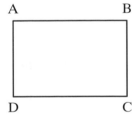

6 Get Real!

The next-door neighbour has a garden exactly the same size.

He plants a tree exactly in the middle of the garden.

a Construct a plan of the garden and mark the position of the tree.

b He grows a lawn in the garden, but he does not want grass within 2 m of the tree. He also wants a flowerbed, 1 m wide, along edges BC and CD. Shade in the area where he wants the lawn to be.

7 **a** Construct a right-angled triangle ABC, where angle A = 90°, AB = 8 cm and AC = 11 cm.

b Mark all the points that are the same distance from AB and BC, and closer to B than C.

8 Get Real!

A, B and C are three television transmitters.

A is 30 km due west of B, and C is 24 km due south of B.

a Use a scale of 1 cm to 4 km to make a scale drawing of the transmitters.

b Anyone within 12 km of transmitter B receives their signal from B. Beyond this, they use either transmitter A or C, whichever is the closest.

Construct the area where transmitter A is used.

9 Get Real!

A man devises a brilliant plan to keep his two goats apart, which will still allow them to graze over the widest area of his field, which measures 22 m wide and 30 m long.

He erects two posts, 12 m apart, with a ring on the top of each post.

He passes a 23 m length of rope through the rings, and ties a goat to each end of the rope.

The diagram shows the position of the two posts in the field.

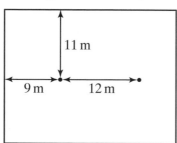

Make a scale drawing of the field, and shade the area the goats can reach.

Homework 2

1 Construct the graphs of these loci.

    a $x^2 + y^2 = 25$ **c** $y^2 = 144 - x^2$

    b $x^2 + y^2 = 1$ **d** $x^2 + y^2 = 13$

2 Write the equation of the circle if A is the point:

a (3, 0)   <u>c</u> (15, 0)

b (7, 0)   <u>d</u> (1.6, 0)

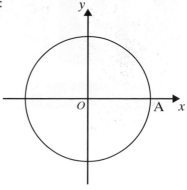

3 Harry says that the graph of $x^2 + y^2 = 25$ passes through the point $(4, -3)$. Is he correct? Give a reason for your answer.

4 Jenny says that the graph of $x^2 + y^2 = 15$ passes through the point $(4, -1)$. Is she correct? Give a reason for your answer.

5 What is the equation of the circle, centre $(0, 0)$, which passes through $(2, -4)$?

6 By drawing graphs, solve these pairs of simultaneous equations.

a $x^2 + y^2 = 16$ and $y = x - 4$ d $x^2 + y^2 = 64$ and $y = -8$

b $x^2 + y^2 = 36$ and $y = x + 1$ e $x^2 + y^2 = 81$ and $x = 9$

c $x^2 + y^2 = 1$ and $y = \frac{1}{2}x - 1$ <u>f</u> $x^2 + y^2 = 17$ and $y = 2 - x$

<u>7</u> A circle, centre the origin, is crossed by a straight line at $(1, 2)$.

a Write the equation of the circle.

b If the circle and the straight line also meet at $(-2, -1)$, what is the equation of the straight line?

Homework 1

1 **a** Sketch these graphs when translated by $\begin{pmatrix} 0 \\ -2 \end{pmatrix}$.

    i $y = x$ **iii** $y = x^3$ **v** $y = \cos x$

    ii $y = x^2$ **iv** $y = \sin x$

  b Sketch these graphs when translated by $\begin{pmatrix} -3 \\ 0 \end{pmatrix}$.

    i $y = x$ **ii** $y = x^2$ **iii** $y = x^3$

  c Sketch these graphs when stretched by a scale factor of 2 parallel to the y-axis.

    i $y = x$ **iii** $y = x^3$ **v** $y = \cos x$

    ii $y = x^2$ **iv** $y = \sin x$

  d Sketch these graphs when stretched by a scale factor of $\frac{1}{2}$ parallel to the x-axis.

    i $y = x$ **iii** $y = x^3$ **v** $y = \cos x$

    ii $y = x^2$ **iv** $y = \sin x$

  e Write the equations for all your sketches.

2 **a** Match each graph with the correct equation.

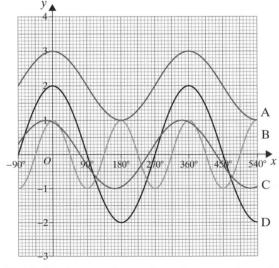

I	$y = \cos(x + 30°)$
II	$y = 2 \cos x$
III	$y = \cos 2x$
IV	$y = \cos x + 2$

  b Describe the transformation of $y = \cos x$ in each case.

3 **a** Sketch the graph of $y = \sin x$ for $0 \leqslant x \leqslant 360°$

b Using this graph, sketch the graphs of:

    i $y = \sin x + 3$ **iii** $y = \sin(x - 45°)$ **v** $y = \sin 3x$

    ii $y = 3\sin x$ **iv** $y = \sin(x + 45°)$ **vi** $y = \sin \dfrac{x}{3}$

4 This is the graph of $y = x^3 - 2x^2$

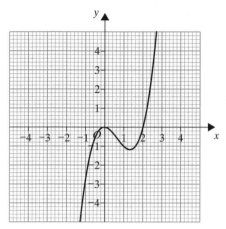

Sketch the graphs of:

a $y = x^3 - 2x^2 + 2$ **b** $y = x^3 - 2x^2 - 2$ **c** $y = 3x^3 - 6x^2$

5 This is Jessie's sketch of the graph of $y = (x + 2)^3$

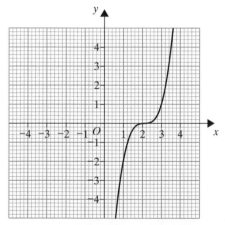

Is she correct? Give a reason for your answer.

6   This is the graph of $y = x^2 - 5x + 4$

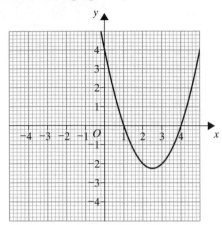

Sketch, in the range $-4 \leqslant x \leqslant 4$, the graphs:

a $y = (x + 3)^2 - 5(x + 3) + 4$

b $y = x^2 - 5x + 6$

c $y = 2x^2 - 10x + 8$

7   Imran thinks that if you translate the graph of $y = x^3$ by $\begin{pmatrix} -3 \\ -3 \end{pmatrix}$ then the resulting graph will have the equation $y = (x - 3)^3 - 3$

Is he correct? Give a reason for your answer.

8   This is the graph of $y = x^2 + 4x$

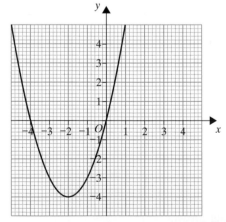

a i   Translate the graph of $y = x^2 + 4x$ by $\begin{pmatrix} 0 \\ 2 \end{pmatrix}$.

ii   Find the equation of the new graph in the form $y = ax^2 + bx + c$

iii Check your answer using a graphic calculator or graphing software.

b i Translate the graph of $y = x^2 + 4x$ by $\begin{pmatrix} 0 \\ -3 \end{pmatrix}$.

   ii Find the equation of the new graph in the form $y = ax^2 + bx + c$

   iii Check your answer using a graphic calculator or graphing software.

c i Translate the graph of $y = x^2 + 4x$ by $\begin{pmatrix} 4 \\ 0 \end{pmatrix}$.

   ii Find the equation of the new graph in the form $y = ax^2 + bx + c$

   iii Check your answer using a graphic calculator or graphing software.

d i Translate the graph of $y = x^2 + 4x$ by $\begin{pmatrix} -3 \\ 0 \end{pmatrix}$.

   ii Find the equation of the new graph in the form $y = ax^2 + bx + c$

   iii Check your answer using a graphic calculator or graphing software.

e i Translate the graph of $y = x^2 + 4x$ by $\begin{pmatrix} 1 \\ 2 \end{pmatrix}$.

   ii Find the equation of the new graph in the form $y = ax^2 + bx + c$

   iii Check your answer using a graphic calculator or graphing software.

9   **a** Explain why if $f(x) = x^2$, the graph of $y = f(x)$ and $y = f(-x)$ are the same.

   b Find another function that has the property $f(x) = f(-x)$

10   **a** Mary says 'it is easy to draw the graph of $y = x^2 + 2x + 1$'. Her method is to translate the graph of $y = x^2$ by $\begin{pmatrix} -1 \\ 0 \end{pmatrix}$.

   Explain why Mary's method works.

   b Billy likes Mary's method. He realises he can draw the graph of $y = x^2 - 4x + 6$ by translating the graph of $y = x^2$ by $\begin{pmatrix} 2 \\ 2 \end{pmatrix}$.

   Explain why Billy's method works.

   c Find the vector that translates $y = x^2$ onto the graph of $y = x^2 - 6x + 5$

11 This is the graph of $y = f(x)$

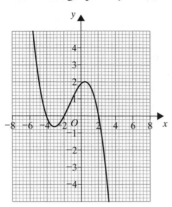

a Match each of the following six graphs with the correct equation.

I	$y = f(x-2)$	**III**	$y = f(x) + 2$	**V**	$y = 2f(x)$
II	$y = f(2x)$	**IV**	$y = f(x+2)$	**VI**	$y = f(x) - 2$

A

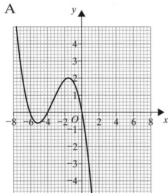

B

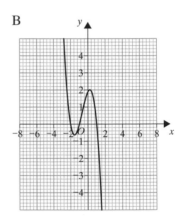

C

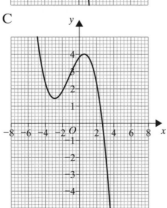

D

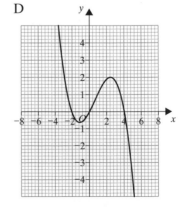

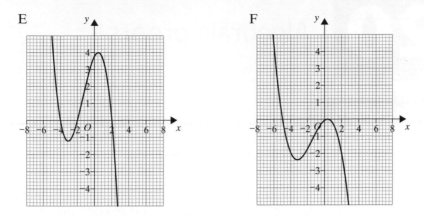

b Describe the transformation in each case.

20 Algebraic proofs

Homework 1

1   Is the value of $3n + 1$ always an even number?

Give a reason for your answer.

2   If q is an even number explain why $(q - 1)(q + 1)$ is an odd number.

3   P is a prime number.

Q is an even number.

State whether each of the following is:

i always even

ii always odd

iii could be either even or odd.

   a PQ

   b $P(Q + 1)$

4   Dewi says that the sum of two prime numbers is always even.

Give a counter example to show that Dewi is wrong.

5   Sue says that the square root of a number is always smaller than the number itself.

Is Sue correct?

Give a reason for your answer.

6   Pam says that the cube of a number is always bigger than the number itself.

Is Pam correct?

Give a reason for your answer.

7   Prove that the sum of three consecutive numbers is always a multiple of three.

8   Prove that the sum of two consecutive odd numbers is always an even number.

9 Prove that the product of any two consecutive even numbers is always a multiple of 4.

10 Andrew says that the difference between two prime numbers is always an even number.

Is Andrew correct?

Give a reason for your answer.

11 Margaret says that if n is an odd number then $4n - 1$ is a prime number.

Is Margaret correct?

Give a reason for your answer.

12 Part of a number grid is shown below:

1	2	3	4	5	6	7
8	9	10	11	12	13	14
15	16	17	18	19	20	21
22	23	24	25	26	27	28
29	30	31	32	33	34	35
36	37	38	39	40	41	42
43	44	45	46	47	48	49

The shaded cross is called C_{10} because it has 10 in the centre.

a This is C_n

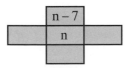

Copy and fill in the empty boxes on C_n

b Neeta notices that $\quad 3 + 11 + 17 + 9 = 40$

and that $\qquad\qquad\qquad 4 \times 10 = 40$

Show, using algebra, that the sum of the arms of any cross is equal to four times the number in the centre of the cross.

13 Prove that the product of two odd numbers is always an odd number.

14 Prove that, for any positive integer n, $n^3 + n$ is always even.

15 The nth term of a sequence is given by the formula $\frac{1}{2}n(n - 1)$

Prove that the sum of any two consecutive terms is always a square number.

Using and applying mathematics coursework task

As part of your GCSE course you are required to submit two coursework tasks covering:

- Using and applying mathematics
- Handling data

Each coursework task is worth 10% of the total marks for your GCSE so it is important that you spend time on each piece of coursework and try to get as high a mark as possible.

This chapter explains how to get the best marks in the Using and applying mathematics coursework task. The Handling data coursework task is discussed in Homework Book 1.

Marks are awarded for the coursework task under three headings, called strands. Each strand is marked out of 8 marks. At the higher tier, you should aim to get between 5 and 8 marks.

Strand 1   Making and monitoring decisions to solve problems

This strand is about deciding what needs to be done and then doing it.
You need to select an appropriate approach, find information and
introduce questions of your own that develop the task further.
For the higher marks you need to analyse alternative mathematical
approaches and make use of higher tier mathematical content.

Mark	Making and monitoring decisions
3	You must consider the given task and obtain the necessary information to solve it.
4	Break down the given task and solve the task in an orderly manner.
5	Introduce your own *relevant* question(s) to develop the task beyond the given task.
6	Develop and follow through alternative approaches making use of more demanding mathematics.
7	Analyse and give reasons for your alternative approaches, making use of a number of mathematical variables or features.
8	Explore independently and extensively a range of appropriate mathematical approaches to the task.

At the higher tier, you should aim to get between 5 and 8 marks.

Strand 2 Communicating mathematically

This strand is about communicating what you are doing using words, tables, diagrams, graphs and symbols (algebra). Your chosen presentation should be used to identify patterns and provide generalisations.

For the higher marks you will need to make use of higher level algebra accurately, concisely and efficiently in presenting a well reasoned argument.

Mark	Communicating mathematically
3	Collect together your information using tables, diagrams, graphs or symbols (algebra).
4	Use *appropriate* presentation along with linking commentary and interpretation.
5	Make use of algebra in forming generalisations and using substitution to check them.
6	Use algebra accurately to support the work and provide justifications.
7	Use higher level algebra accurately in presenting a reasoned argument for your work.
8	Use higher level algebra concisely and efficiently in presenting a reasoned argument for your work.

At the higher tier, you should aim to get between 5 and 8 marks.

Strand 3   Developing skills of mathematical reasoning

This strand requires you to search for patterns and provide generalisations for your task. Generalisations should then be tested on new data, justified and explained.

For the higher marks you will need to provide a sophisticated and rigorous justification, argument or proof which demonstrates mathematical insight into the problem.

Mark	Developing skills of mathematical reasoning
3	Use the information collected to make a generalisation that is true for all the results.
4	Check the generalisation by testing a further example involving new data.
5	Provide a justification for the generalisation; the justification can be algebraic, graphical or diagrammatic.
6	Draw together the generalisations for the extended task and provide a justification for these.
7	Provide an overarching justification that coordinates a number of mathematical variables or features.
8	Provide a mathematically rigorous justification, argument or proof that includes the conditions for its validity.

At the higher tier, you should aim to get between 5 and 8 marks.

Explore 1 Number grid

Look at this number grid:

1	2	3	4	5	6	7	8	9	10
11	12	13	14	15	16	17	18	19	20
21	22	23	24	25	26	27	28	29	30
31	32	33	34	35	36	37	38	39	40
41	42	43	44	45	46	47	48	49	50
51	52	53	55	55	56	57	58	59	60
61	62	63	64	65	66	67	68	69	70
71	72	73	74	75	76	77	78	79	80
81	82	83	84	85	86	87	88	89	90
91	92	93	94	95	96	97	98	99	100

- A box is drawn around four numbers
- Find the product of the top left number and the bottom right number in this box
- Do the same with the top right and bottom left numbers
- Calculate the difference between these products

Investigate further

This task is an AQA set task so it can be submitted for marking by AQA or marked by the centre.

The words 'Investigate further' mean that you should develop the task beyond its original scope.

There are no right or wrong ways in which the task can be developed. Think of different ways in which you might extend the task.

Some tips

You may wish to consider the following:

- Think about how you might extend the task. In what ways might the task be developed? Try out some of these ways and remember to comment on your findings.

- Remember to include tables (or graphs) to illustrate your findings. Use your tables (or graphs) to write down any patterns that you notice. Remember to show all your working.

- Think about your findings and try to explain them. Are you sure that the results are correct? How do you know? Have you tried all of the possible variations?

- Use algebra to develop the task further. Can you write the product in algebraic terms? Can you write the difference in algebraic terms? Make sure you explain what you are doing.

Explore 2 Trays

A shopkeeper asks a company to make some trays.

A net of a tray made from a piece of card measuring 18 cm by 18 cm is shown below:

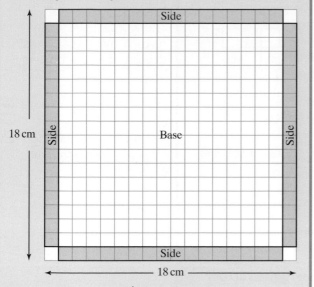

The shopkeeper says, 'When the area of the base is the same as the area of the four sides, the volume of the tray will be a maximum.'

◎ Investigate this claim

Investigate further

This task is an AQA set task so it can be submitted for marking by AQA or marked by the centre.

The words 'Investigate further' mean that you should develop the task beyond its original scope.

There are no right or wrong ways in which the task can be developed. Think of different ways in which you might extend the task.

Some tips

You may wish to consider the following:

◎ Use squared paper to create the net of the tray, then use your net to make the box. What is the area of the base? What is the area of the sides? What is the volume of the tray? What do you notice?

◎ Think about how you might extend the task. In what ways might the task be developed?

◎ Remember to include tables or graphs to illustrate your findings. Use your tables or graphs to write down any patterns that you notice. Remember to show all your working.

◎ Use algebra to develop the task further. Can you write the areas in algebraic terms? Can you write the volumes in algebraic terms? What do you notice? Make sure you explain what you are doing.

Explore 3 Tiles

Assad is making patterns from square tiles.

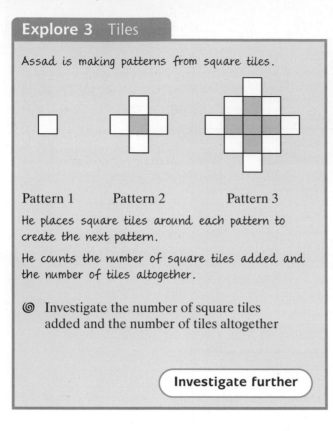

Pattern 1   Pattern 2   Pattern 3

He places square tiles around each pattern to create the next pattern.

He counts the number of square tiles added and the number of tiles altogether.

◎ Investigate the number of square tiles added and the number of tiles altogether

Investigate further

This task is not an AQA set task so must be marked by the centre and moderated by AQA.

The words 'Investigate further' mean that you should develop the task beyond its original scope.

There are no right or wrong ways in which the task can be developed. Think of different ways in which you might extend the task.

Some tips

You may wish to consider the following:

◎ Try out some further patterns and make a note of your findings. Think about your findings and try to explain them. What do you notice about the number of square tiles added and the number of tiles altogether?

◎ Remember to include tables (or graphs) to illustrate your findings. Use your tables (or graphs) to write down any patterns that you notice. Remember to show all your working.

◎ Think about how you might extend the task. In what ways might the task be developed? Try out some of these ways and remember to comment on your findings.

◎ Use algebra to develop the task further. Can you write down the nth term for the number of square tiles added and the nth term for the number of tiles altogether? Make sure you explain what you are doing.